Chasing Monarchs

Books by Robert Michael Pyle

Wintergreen: Rambles in a Ravaged Land

The Thunder Tree: Lessons from an Urban Wildland

Where Bigfoot Walks: Crossing the Dark Divide

Nabokov's Butterflies: Unpublished and Uncollected Writings (with Brian Boyd and Dmitri Nabokov)

Chasing Monarchs: Migrating with the Butterflies of Passage

Walking the High Ridge: Life as Field Trip

Sky Time in Gray's River: Living for Keeps in a Forgotten Place

Mariposa Road: The First Butterfly Big Year

Letting the Flies Out (poems, essays, stories)

The Tangled Bank: Writings from Orion

Evolution of the Genus Iris (poems)

ON ENTOMOLOGY

Watching Washington Butterflies: An Interpretive Guide to the State's 134 Species, Including Most of the Butterflies of Oregon, Idaho and British Columbia

The National Audubon Society Field Guide to North American Butterflies

The IUCN Invertebrate Red Data Book (with Susan M. Wells and N. Mark Collins)

Handbook for Butterfly Watchers

Butterflies: A Peterson Field Guide Color-In Book (with Roger Tory Peterson and Sarah Anne Hughes)

Insects: A Peterson Field Guide Coloring Book (with Kristin Kest)

The Butterflies of Cascadia: A Field Guide to All the Species of Washington, Oregon, and Surrounding Territories

Chasing Monarchs

Migrating with the Butterflies of Passage

Robert Michael Pyle

Foreword by Lincoln P. Brower
New Afterword by the Author

Yale
UNIVERSITY
PRESS
New Haven and London

First Yale University Press edition published 2014.
First edition published by Houghton Mifflin Company 1999.

Yale University Press books may be purchased in quantity for educational, business, or promotional use. For information, please e-mail sales.press@yale.edu (US office) or sales@yaleup.co.uk (UK office).

Drawing on title page by Thea Linnaea Pyle.
Book design by Anne Chalmers.
Printed in the United States of America.

Library of Congress Control Number: 2013954244
ISBN 978-0-300-20387-5 (pbk.)

A catalogue record for this book is available from the British Library.

10 9 8 7 6 5 4 3 2 1

For Thea

I feel drawn south again, into wild, fragrant places.

— Vladimir Nabokov, letter to his mother, Elena Nabokov, August 15, 1929

What I wanted was guidance, a system of telemetry to ease the tension of not knowing what happens next.

— Alison Deming, from #54, *The Monarchs: A Poem Sequence*

It is long after sunset when I look up in the twilight and see a monarch flying overhead, a small, black, fluttering form against the fading glow in the sky. It is headed south.

— Edwin Way Teale, *A Walk Through the Year*

But most of all I shall remember the Monarchs, that unhurried westward drift of one small winged form after another, each drawn by some invisible force. . . . Did they return? We thought not; for most at least, this was the closing journey of their lives.

— Rachel Carson, letter to Dorothy Freeman, September 1963

Contents

Foreword by Lincoln P. Brower · ix

Maps · xii

Beginnings: Sky River · 1

ONE Similkameen · 11

TWO Okanogan · 28

THREE Grand Coulee · 43

FOUR Crab · 57

FIVE Columbia · 73

SIX Deschutes–John Day · 101

SEVEN Hell's Canyon · 118

EIGHT Snake · 127

NINE Bear · 145

TEN Bonneville · 159

ELEVEN Apache Gold · 181

TWELVE Guadalupe · 197

THIRTEEN Buenos Aires · 210

FOURTEEN Cibola · 223

FIFTEEN Pacifico · 236

SIXTEEN Fandango · 253

Endings: Recovery · 268

Afterword · 275

APPENDIX *Conserving the Monarch of the Americas* · 285

Further Reading and Resources · 289

Acknowledgments · 299

Index · 303

Foreword

Lincoln P. Brower

I am delighted to welcome the new release of *Chasing Monarchs* by my good friend and master lepidopterist, Robert Michael Pyle. We have long shared an appreciation for the monarch butterfly and its complex migration biology. We have also long shared a deep concern for its conservation, and have worked together toward that end, including memorable travels in Mexico. Since Bob and I were both graduate students of the late legendary lepidopterist and professor of biology Charles Remington at Yale, I am particularly pleased that *Chasing Monarchs* is coming back into print with Yale University Press.

In the lyrical and engaging *Chasing Monarchs,* Pyle recounts his 1996 journey with monarchs as they migrated across the northwestern United States, southward through the Great Basin to the Rocky Mountains and finally to the Arizona border with Mexico. Pyle's expedition was inspired by his questioning of what he calls the "Berlin Wall" model of monarch migration. The general consensus by the 1990s was that all monarchs that breed east of the Rocky Mountains migrate to overwintering sites in the Oyamel fir forests in central Mexico, whereas monarchs breeding in the western United States migrate to overwintering sites along the California coast. Accordingly, the Rocky Mountains were seen as a dividing line between what were assumed to be completely distinct eastern and western breeding populations. Gradually accumulating evidence did not always agree with this dogma, and Pyle—the consummate naturalist—decided to follow individual monarchs and see for himself what was really happening in western North America. *Chasing Monarchs* tells of his nine-thousand-mile journey of discovery.

On his expedition, Pyle recorded the direction of flight of as many monarchs as he could find. As he had predicted, a substantial number of the sixty-two individuals showing directional flight were headed not toward the California coastal areas but toward the south and southeast. Several research papers published since the original publication of *Chasing Monarchs* support the hypothesis that the eastern and western populations are not completely separated. Molecular genetic studies by several investigators have found little genetic differentiation between the eastern and western populations. Definitive evidence of some degree of population mixing comes from the Southwest Monarch Study, which has tagged a large number of monarchs in Arizona since 2003. Arizona butterflies have been recaptured at overwintering sites both in California and in Michoacán, Mexico.

The emerging evidence that some monarchs that breed west of the Rocky Mountains do migrate to central Mexico raises a new series of intriguing questions. The most obvious is, how frequent is the interchange of butterflies between the two populations, and is the movement primarily in one direction, or in both directions? Do the western adults that winter in Mexico remigrate to the West, or do some of them fly into Texas and mix with the eastern North American migrants? Likewise, how often do eastern migrants get caught in winds that carry them into the western range, as Pyle and I suggested in a coauthored paper in 2004? Has what Pyle called the "Berlin Wall" model been dismantled, or does it have a few cracks? Once these questions have been answered, a plethora of genetic, evolutionary, and ecological questions can be addressed.

I hope that there will be sufficient time, and sufficiently large monarch populations, for scientists to pursue these investigations. In 1980 Pyle and I recognized that the monarch butterfly migration and overwintering behavior had become an endangered biological phenomenon. Degradation of the Oyamel forest was threatening the overwintering enclaves in Mexico, and real estate development in California was destroying several of the west coast wintering sites. Both Pyle and I were also concerned about the increasing use of agricultural herbicides, which exert massive collateral damage by killing the butterflies' nectar sources as well as the milkweeds that are the only plants that monarch caterpillars are able to eat. In the decades since, while the popularity of the monarch butterfly has soared, its numbers have not.

Annual censuses in Mexico and California have documented precipitous declines in the number of monarch butterflies wintering in both populations, as reported in recent papers by Brower and colleagues and from the Xerces Society. The total area occupied by monarchs in Mexico declined to an all-time low in the 2012–2013 overwintering season. The dwindling of the eastern numbers has undoubtedly been accelerated by the blanket deployment of genetically engineered corn and soybean strains that are completely resistant to powerful herbicides, leading to the elimination of milkweeds and nectar sources from millions of acres of breeding habitat. These chemicals have caused a 50 percent decline in milkweed in agricultural fields in the Midwestern United States and an 80 percent decline in monarch reproduction in the same area, according to recent evidence from John Pleasants and Karen Oberhauser. Sarina Jepsen and Scott Black of the Xerces Society ascribe the decline in the number of monarch butterflies wintering in California at least partially to a degradation of western breeding and wintering habitats. And if all this were not enough, chronic drought and climatic warming may exacerbate the problems faced by both Californian and Mexican wintering monarchs.

Writing in *Orion* magazine in 2001, I referred to the monarch as the canary in the cornfield. It now appears to be gasping, and conservation action on a broad scale is needed more than ever. At a recent gathering at Princeton University, I was asked why the loss of the migratory monarchs would matter. I said, after a slow deep breath, "Well, what difference would it make if we lost the *Mona Lisa?*"

Bob Pyle is an eloquent champion of the monarch butterfly. Readers who join him on his *Chasing Monarchs* journey will understand why it is so important to conserve this magnificent butterfly and its extraordinary migration.

The Monarch Chase

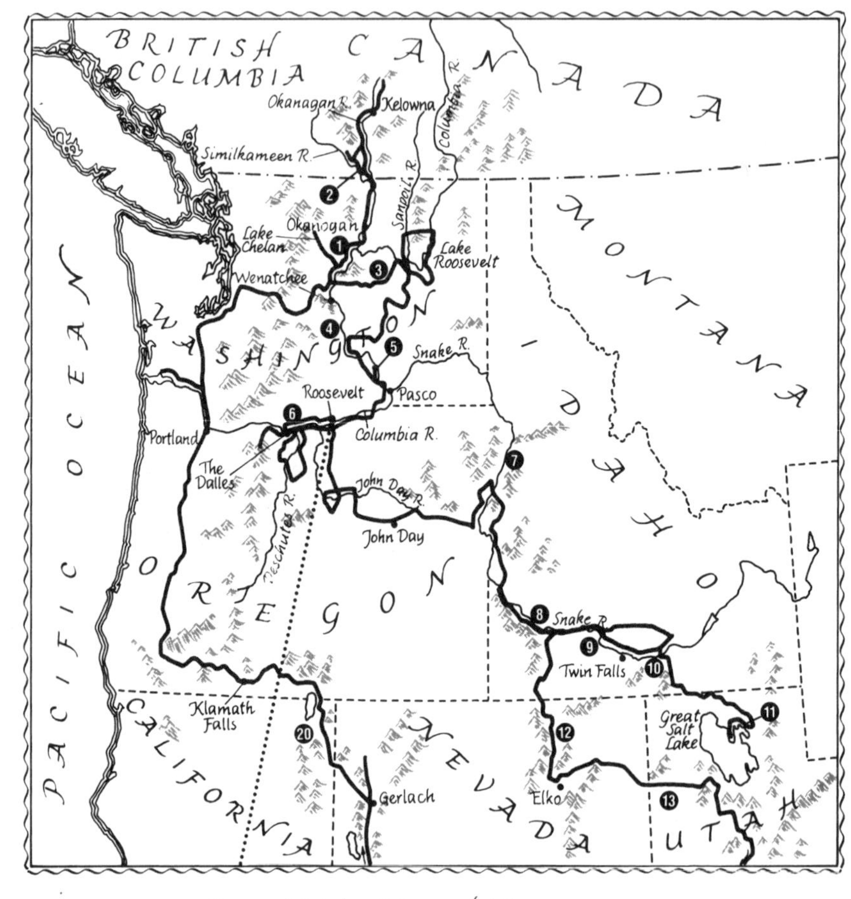

Sites of Major Monarch Encounters

1. Gallagher Flat
2. Cawston
3. Sun Lakes
4. Crab Creek
5. Hanford Reach
6. Maryhill
7. Hell's Canyon
8. C.J. Strike Dam
9. Thousand Springs
10. Burley
11. Bear River Refuge
12. Owyhee Grade
13. Bonneville Salt Flats

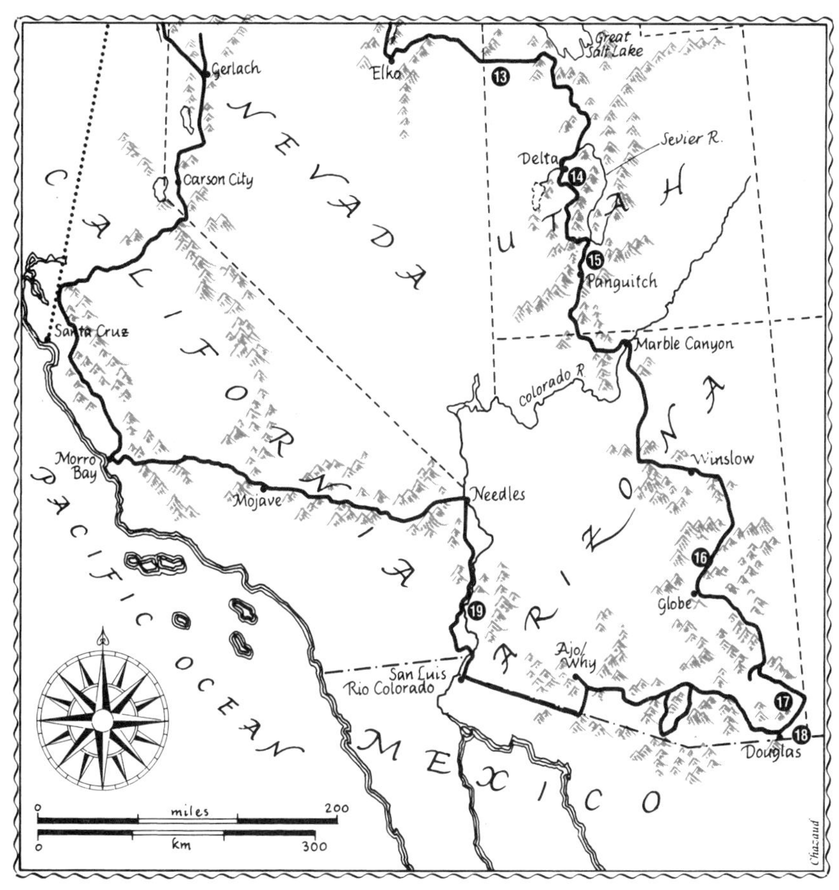

Route starts and ends on map at left

⓭ Bonneville Salt Flats

⓮ Clear Lake Refuge

⓯ Sevier Canyon

⓰ Salt River Canyon

⓱ Chiricahua Mountains

⓲ Guadalupe Canyon

⓳ Cibola Refuge

⓴ Fandango Pass

——— *Author's route*

❶ *Monarch sightings*

·········· *Flight path of Monarch #09727*

Chasing Monarchs

BEGINNINGS

Sky River

No ORANGE SHOWS *in the tall firs. In the predawn pallor, the bark is cloaked with a gray nap. The boughs look swollen, as if hung with Spanish moss. The gray burden is butterflies, their wings folded to show only the dull undersides. Then the sky breaks open to the east and tentative sunbeams fondle the higher boughs. A rustle begins, almost too soft to hear, as the temperature rises. A wing spreads, then another. The orange appears. A daub, then a streak; slowly spreading at first, a cool fire in the canopy becomes a conflagration in minutes as a million butterflies lift off. Now orange fills the forest, eclipsing the green beneath the high blue Mexican sky.*

The monarch butterflies have hung here, in the mountains of Michoacán, since late fall. Mostly torpid, they have gone out only to soar on hot days when a shadow alarmed them, to drink at the creekside below the forest, and (for those that didn't fatten enough for the winter) to seek nectar among sparse salvias and senecios. Now it is early spring, and the butterflies' sex urge has awakened. Their compasses, formerly set for south, have been recalibrated. Today, after dropping out of the trees, some will not return to the clusters as the afternoon cools. Tomorrow more will depart. The freezes are past, and milkweed is sprouting in the North. It is time to go.

Now, its broad, tawny vanes expanded to the sun and shivering to warm up even faster, the first monarch of many millions drops over the edge of the scarp below its high-country wintering place. Others follow. Soon the valleys that funneled the monarchs up the montane corridors last autumn will pour their molten flow back into the lowlands. Exodus becomes diaspora, as emigration turns toward reinhabitation.

In October these milkweed butterflies comprised a torrent flowing southward, sometimes braiding into many narrower streams, sometimes congealing into a broad front constantly reinforced by tributaries. At night they wrapped themselves in trees, previews of the dense masses they would make in their winter woods farther south. They roosted in sumacs, blending into reds and purples deeper than their own dusk colors. They roosted in locusts, clinging in the wind as tenaciously as the trees' own clusters of pods, and in basswood, elm, and oak. They roosted in osage orange, mulberry, and pecan; in willow, cottonwood, and mesquite, each night growing heavier in number. But now, drifting north, there is little togetherness; it's every monarch for itself. All the males want to do is travel and find females; all the females want to do is find milkweed plants — the sole acceptable food for their offspring — and stay on the move. At night they roost alone or with a mate at most.

Most of the females will have mated before leaving the mountain or shortly into their journey. With a continent to cover, there is no time for the elaborate courtship many butterflies perform. The forest colony in evacuation resembles an orgy. Spotting a darker-veined, russet female, the paler male pounces in midair. He attempts to ride, to skyjack the usually smaller female, often taking her to the ground. There, among the needles and duff, he forces her onto her back if he can, and immediately copulates with her. Then he carries her into a tree in a postnuptial flight.

Females seeking milkweed, males seeking females, both seeking nectar, the monarchs fly north until they die along the way. Their wings, those translucent tissues that have carried them two hundred or two thousand miles already, lose their scales and their color. They grow tattered, torn, bush-ripped, bird-struck. The wings still work for flying, if not so fast, efficiently, or high. But a female no longer needs to fly high, to soar and glide great distances. Her only need now is to skip and hop and flutter from field to roadside, bush to herb, to palp the air and the green spring for the next scrap of milkweed on which to deposit her precious eggs. Day after day she drops her load until she has laid some five hundred eggs or more. Or until a spring storm or an automobile or a recently molted mantis lays her to rest. Or until, her abdomen grown greasy with the thinness of its walls, her wings pale rags, she simply wears out.

A few monarchs may reach the state, the county, the meadow where they began, but this must be very rare. The great mass of last fall's emigrants scatter their last scales to the wind somewhere far short of their birthplaces. It will be up to their offspring to carry on. The eggs — fluted golden ovals — darken in a few days, then give forth comma-sized caterpillars. These consume milkweed flesh until they burst their skins, five times. As they gnaw the host plant, they spill its latex, which will embitter their own tissues. In less than a month the caterpillars have grown as large as a child's little finger. After the fifth molt comes the jade jewelbox, the gold-nippled smooth green chrysalis. From this inch-long pellet will come a butterfly whose wings span three or four inches. Once they break free of their confining chrysalides, then dry in full turgor, they are capable of rising, soaring, and gliding better than any kite. On the first morning breeze they take wing, rise on the thermals, level out, float and pump higher, higher, until, hundreds or perhaps thousands of feet in the air, they begin the masterful glide that can carry them many miles. If the air is heavy, they move close to the earth. And so from the millions of shimmering pupae, tucked beneath the leaves of milkweed from St. Augustine to Santa Barbara, will issue the next rush of the sky river, flowing north and east.

Even before the mountain forests deep in Mexico give up their monarchs, a similar turnout takes place along the California coast, Alta and Baja. As the offspring of the Mexican emigrants emerge, poised to flow north across the south and center of the continent, those of the West Coast overwinterers are dispersing as well, all up and down the Coast Ranges, California's Central Valley, and the Sierra Nevada foothills.

As the air stills in the afternoon, the voyageurs drop into likely territory to find nectar and mates and milkweed again, until dusk or cold wind forces them to take shelter. If bad weather sets in, they may be grounded for days. But as soon as the sun and wind and rain allow, they forge on into the mountains, valleys, and canyons of the north — wherever milkweed grows and sometimes far beyond, for each monarch will continue the search until it drops.

Where rivers, creeks, canals, or roadsides go generally their way, the wanderers (as monarchs are also called) follow them. Here is water, here is nectar, here the likelihood of the sought-after herbs. Water-courses deliver them to milkweed grounds: fields, pastures, meadows, banks and bars; roads lead to parks and gardens and vacant lots, where few milkweed plants are likely to go undiscovered by *Danaus* in a good year.

This is how the sky river floods North America each spring. Wave upon wave of orange insects, whose grandparents ebbed out of their summer range the previous autumn, now pour in. Through each declivity in the landscape — arroyo, ravine, gully, valley, canyon, swale — they surge. Eddies occur as the sky river gives off another generation and then another, each of which will live just weeks. Then, having covered three countries to the extent of the milkweed, the last riffles probe still farther, seeking an ever broader beachhead for the species. So by September, in years when conditions are right, the continent once again brims with butterflies inclined to leave: the new emigrants, three or four generations removed from those of last fall. Then the current turns back again, to become not a meandering, chaotic mob like their northbound ancestors but a single-minded, gathering onslaught on the southern reaches.

So the sky river runs two ways, north with the spring, south with the fall. First the millions of individual rivulets flow south from the continental borders to the core flyways, and on down to the great central pools. Then, from those winter reservoirs, the flow reverses to refill the hinterlands. This is a current that changes direction with the season, a tide with a turnaround time of six months instead of six hours. Or a breath, the lungs of the land respiring butterflies out into the far capillaries of the continent, then sucking them back again, a great sigh, in by winter, out by summer.

Pick your metaphor as you will. To the monarchs it is just what they do. No matter to them that their regular, stylized shifts of population comprise the grandest butterfly spectacle on the planet. It is in the nature of all other animals but humans to carry on heedless of any glamour, show, or drama they might possess. We are not so oblivious. Confronted with thousands or millions of big, flapping, soaring, gliding, sucking creatures clearly going somewhere, we stand in awe. We want to know, Where have they come from? Where are they going? How? We want to know what's going on here.

"So, just how *do* you follow butterflies?"

I was asked that question again and again when I mentioned my plans to follow monarchs. It surprised me a little. But then, I had to remind myself, the people who asked had not been following butterflies for most of their lives, as I had.

But had I, really? It's true, I'd been chasing butterflies for nearly forty years. But most of the pursuits had been decided within a few yards, or at most a field or two. Lepidopterists over the age of twelve seldom outright *chase* a butterfly. Sure, occasionally they'll follow one over a meadow or across a stream, and with luck they might even catch it, but once a butterfly decides to go, it's gone.

In recent years I've done a lot more watching than catching. I've often observed a nectaring or courting butterfly for half an hour or more. Not waving a net, I present a less threatening vision, and I'm often able to move along with the butterfly and keep on its good side. The careful stalk, sometimes still, sometimes liquidly in motion, over and over, is the stock in trade of all hunters. That certainly applies to those who hunt butterflies, for there never was a more now-you-see-it, now-you-don't kind of prey.

To people's questions, I replied: I'll find a monarch. I will watch it. If it flies, I'll follow it as far as I can. When I lose it, I'll take its vanishing bearing — the direction in which it disappears. Then I will quarter the countryside, by foot and by road, until I find the next suitable habitat along that bearing, and do it again. Monarch by monarch, shouldn't I be able to follow their trend?

Friends also wondered whether I would use high technology. Will you strap microtransmitters on the monarchs? they'd ask in jest. In fact, such devices do now exist for honeybees and ground beetles. In Alberta, tiny dipole antennae thinner than a hair, attached to minute microwave-echoing diodes, have been affixed to parnassian butterflies and tent caterpillars and their moths. But the tenth-of-a-milligram tag and its radar receiver unit have a range of only forty-five meters, and you can easily follow a monarch that far by sight.

Others wanted to know if I would fly an ultralight plane, "like the guy with the geese." No, I'd reply. I'm a low-tech kind of a guy. It will be me; my trusty butterfly net, Marsha; my old Honda, Powdermilk; boots; a little beer; some knowledge of what butterflies do and how to find them; and a strong dose of curiosity mixed with a passion to know where monarchs fly, and how.

Still, I had to ask myself whether my prior experience had prepared me for following a butterfly — in particular *this* butterfly, the most robust and far-flung of all — over a long distance. To me, the more interesting question wasn't *how* I was going to follow them but *why.* Faced with any number of ways to spend one's autumn profitably, or at least pleasurably, why was I going off on a wild butterfly chase?

The monarch, *Danaus plexippus* (Linnaeus), also known as the milkweed butterfly and the wanderer, conducts probably the longest and certainly the most spectacular mass movement of any insect. Most butterflies that reside in temperate climes get through the cold season by entering a phase known as winter diapause. An antifreeze-like substance in the animal's tissues keeps it alive during the diapause, which may take place in the egg, larval, or pupal stage, depending on the species. A few kinds, such as the mourning cloak and its relatives, even diapause as adults by hibernating in a hollow tree. Insects lacking a winter diapause phase must live in warmer latitudes or else migrate. Most of those that migrate north in the summer, such as painted ladies, perish in the fall rather than return south.

Many butterflies engage in mass movements to expand their range or relieve pressure on their habitat. Some are on their way to acquiring a diapause or a back-and-forth migration. But in North America only the monarch regularly shifts north in the spring, south in the fall, en masse, like birds.

Monarchs belong to a subfamily, the Danainae, all of whose members are tropical or subtropical and none of which has acquired the ability to overwinter north of the line of regular frosts. Yet their host plants, milkweeds (*Asclepias* species), occur abundantly in North America. To take advantage of that resource, monarchs have evolved their two-way flow, northward or inland in the spring and southward or coastward in the fall. This essential fact seems well established. But for all their popularity and widespread lore, monarchs are far from well known. Large questions remain. What does a monarch's travel itinerary actually look like? It is certainly no straight line, as the vectors on monarch migration maps tend to suggest.

For at least a century, wanderers have been observed spending winters along the California coast — not in particularly warm places but in clumps and bands of trees in the fog belt. Only since 1975 has it been generally known outside of Michoacán that the bulk of the North American monarchs overwinter in Mexico — not in Veracruz or Cancún or any other hot resort where gringos and warblers might go to winter, but on a few high, cool, forested slopes of the transvolcanic ranges west of Mexico City. This long-distance shift of a whole continental population of insects is one of the most astonishing migrations on earth, and the greatest butterfly gathering. Yet it takes place in a profound cloud of mystery.

Three questions stand out for me. First, how do the monarchs physically *do* it, this migration over thousands of miles of North America; what are the lives of the migrants like? Second, do they actually navigate or just ride the wind? And third, why do some monarchs end up in Mexico and others in California, and what determines which parts of the breeding population will reach either destination?

Many aspects of the first riddle may be approachable only at the genetic, cellular, biochemical, and experimental-behavioral levels. But as for how they comport themselves day by day during their long journeys, surely much would be discernible by a careful watcher on the ground. Many naturalists have contributed their observations to a collective picture of monarch behavior. But most of them watch the butterflies go by in the fall and come back in the spring. As far as I knew, no one had ever tried to follow them all along the way.

As for navigation, the presence or absence of a sun compass in monarchs has been the subject of experimentation lately, and studies of their possible orientation by magnetic fields closely followed. Again, a watcher on the ground should be able to get a sense of how the migrants relate to topographical features and whether they seem to select their direction or allow the elements to direct their route.

The third puzzle, of origins and destinations, has been approached by dedicated folks tagging monarchs for more than half a century, ever since Professor Fred Urquhart, the dean of monarch studies, first began affixing sticky labels to their wings and letting them go in hopes they might be found again. Indeed, enough *were* found over the years — some low thousands out of a million or more tagged — to suggest that the eastern monarchs were pointing toward Mexico in their fall exodus. Eventually, in 1975, when Urquhart's associates located the winter clusters in Michoacán, this idea was borne out.

Western monarchs are far less abundant than the eastern, and far fewer have been tagged, especially outside of California. Because it was easy to collect and tag individuals from the wintering trees on the coast, Urquhart and his helpers transplanted California monarchs to other states and provinces to see what would happen. Transplants, of course, cannot demonstrate what monarchs of local origin would do. But because some of these made it back to the West Coast and were recovered there, the unwarranted assumption arose that all western monarchs migrate to the California shore.

Since then, largely due to Urquhart's published maps of tag recoveries, the idea that monarchs breeding east of the Rocky Mountains all go to Mexico, while those breeding west of the Rockies all go to the California coast, has become canon. This belief has been perpetuated by most people who have written about monarchs, and most published maps show the Rockies as a monolithic butterfly barrier. Yet as a boy chasing alpine butterflies high in the Rockies each summer,

I saw monarchs crossing passes — Loveland, Corona — through the Continental Divide in both directions! So I've always been uncomfortable with the received wisdom that East and West shall never meet in the person of monarchs. After making repeated ads, queries, and interviews, and searching the literature, Web, and grapevine, I unearthed only two sources of wild, indigenous monarchs tagged outside California and recovered on the West Coast, both from Boise. Also, two monarchs from the same site in Idaho were recovered far to the southeast in Utah. A basic assumption of North American natural history thus seemed based on thin evidence indeed.

Twenty years ago, when I became deeply involved with efforts to conserve the vast overwintering aggregations of monarchs in Mexico and California, I began to realize that without knowing just how these butterflies spend their days and nights in passage, we could not know how best to provide for their needs along the way. Lincoln Pierson Brower, who has contributed much of our modern knowledge about monarchs, long ago impressed on me the critical importance of what goes on "along the way." As Distinguished Service Professor of Zoology at the University of Florida, he has, along with orchestrating a brilliant research program, devoted much of his energy toward monarch conservation. He realized that their survival does not lie only in protecting the winter roosts or the summer breeding grounds, that the migratory *system* must be taken into account. Yet if we didn't know where the two winter populations came from and how they got where they were going, how could we plan for their future? It would be like trying to conserve a river whose source lies in perpetual cloud.

In spite of all the watchers and taggers over the years, we still lack a daily flight plan for moving monarchs. Monarch Watch, of Lawrence, Kansas, and Journey North use the Internet to connect classrooms and field observers and plot arrivals at the speed of light, mostly from the Great Plains eastward. The Monarch Program, in San Diego, monitors activity chiefly along the California coast. A barren of ignorance lies in between. By physically trailing the beasts through that barren, migrating *with* the monarchs, day by day, north to south, I hoped to shed a few lumens of light on the several mysteries of the monarchs.

This is not a comprehensive treatise on *Danaus plexippus.* While my observations may help to expand our knowledge of this complex and surprising animal, this book is largely anecdotal. It is the story of one person's journey, wandering with the wanderers, allowing another species to be his guide day after day, across great distance.

And so I began in the autumn, when the orange was in the North. The butterflies of passage would lead, and I would follow.

ONE

SIMILKAMEEN

It was the time of asters. Purple asters, mauve asters, tiny white asters. Summer was slipping out of the North, and with it the creatures of passage. On the last day of August, somewhere, monarch butterflies were on the move.

Leaving western Washington for eastern, I noticed that the maritime greens had gone tired and blanched, ready for refreshment from the winter rains. Up on Swauk Pass, the first yellow among the serviceberry leaves rumored frost. That was where I had first seen and photographed a migrant monarch in Washington, twenty-five years before, as it sought nectar in a broad meadow near Blewett Summit. This afternoon the only butterflies on the wing were pine whites, flimsy pale drifters that fly at the end of summer and into the time of asters. Lake Chelan, the ancient glacial fjord in the heart of Washington's highlands, had gone gray from a light wind with the cold bit of autumn in its mouth. Dropping downlake out of the North Cascades, that wind suggested that any sensible butterfly would want to head south soon, if it had the means.

But the next morning was bright and warming at Gallagher Flat, north of Chelan. I intended to begin following monarchs in Canada but to first search for signs of them on my way north. I was not alone at the start. My wife, Thea, set out with me, though she would return home before long, leaving me to make the chase solo. It was Thea, some years before, who had first found monarchs here. Chelan is among the richest and best-studied counties in the Northwest for butterflies, and it is not easy to find something new within its borders.

Thea, who used to live in Chelan, knew Gallagher Flat, a former embayment of the Columbia River, as an excellent site for wild asparagus. Noticing milkweed, she looked one fall for monarchs, and found them — the first recorded from the county. What better place to join these dances with monarchs?

The old river channel lies between State Highway 97 on the east and a curving railroad line on the west, both built well up above the level of the frequently flooding bottoms. Across the roadway burbles the broad Columbia, dulled here now by Rocky Reach Dam. As we walked down onto the flat, we were surrounded by milkweed in various stages of maturation, from plants with soft, fuzzy little pods to those already bursting, seeding their down onto the air, as well as a single flower head still in fragrant pink bloom. A big Oregon swallowtail, the official state butterfly of Oregon, nectaring on a weedy purple vetch, rose, circled, and disappeared into an asparagus forest. This cheddar-yellow, black, and blue flier was the only other species likely to be seen here that is as big as a monarch, and it gave me a start. So did bright red grasshoppers. With my search image not yet calibrated, these flashes of color made potent false alarms. Yet when the real thing appeared, I knew it instantly.

Recent years had not been strong for monarchs in the West. So when the first one showed up, around noon, I was both thrilled and relieved. There *would* be monarchs to follow! We were halfway up the length of the crescent-shaped flat on the river side. The monarch, too, nectared on the royal purple vetch, but a sturdy fence ran between us. Since it was my intention to catch, examine, and physically tag every monarch I could, in the slim hope that one or more might be recovered, I needed to get to it. I found a way through the tight-gauge wire, and then the monarch drifted like a feather, floated over the fence, and dropped onto the side I'd just come from. I shimmied through again, worked myself slowly into position for a wing shot, and swung. But I bricked it — one of the more polite collector's terms for missing a butterfly.

The big orange beast powered up to the highway, then drifted back. I lost it against the crazy quilt of fall vegetation. Having noted its penchant for vetch nectar, I headed for a thick, tangled purple swale, and there it was. This was my first good, close look at a living monarch in months, and I savored it before taking a good, clear swing for what seemed a certain catch. But, hair-triggered, I struck short and bricked it again.

The unsympathetic butterfly, a large, fresh male, was really disturbed this time. Before, it had flapped away from the source of menace in a leisurely way and resettled. Now it flew fast, high up toward the railroad tracks, then topped the bluff to the west, and, circling, disappeared into the clear sky at least a hundred yards above the ground. I saw right off the conflict between tagging and following, because in catching, or missing, I could scare them off. This one flew so high I was unable to get even a vanishing bearing, the direction of disappearance that enables observers to plot or follow an animal's movements.

Still, something about missing my very first monarch seemed right. The Indians up and down the ruined Columbia, the Chinook and Yakama, Umatilla and Warm Springs, observed the rite of releasing the First Salmon of the season in thanks to the Salmon People for coming back. (This practice had many variations, and sometimes just its bones were sent back as emissary.) Of course, I was going to release the monarch anyway after I tagged it. But there is a tincture of enforced humility in missing that has to be salutary for a venture such as this. Pride is no friend to the wandering naturalist.

I couldn't blame the miss on my butterfly net. Marsha's shaft is a five-foot cottonwood branch from the High Line Canal in Colorado. My brother Bud first found and used it as a walking staff. When I needed a new net pole one summer, I appropriated it, made a big hoop and bag for it, and named it for a strong friend. Donald Culross Peattie has written that the Omaha made sacred poles from cottonwoods, "the chief tree of the prairies." Marsha is my sacred pole. I have used her for more than twenty years now, and her skin is burnished from sweat, oil, and flesh. Heavier than most butterfly nets, Marsha is a two-handed tool. Though she was crooked, cracked, and liberally bound with duct tape, when I bricked a butterfly, it was almost never her fault.

But even if this one had gotten away without a trace, I followed it anyhow in my mind. I could see it climbing the Entiat Valley into the Chumstick Mountains, down the other side and across the Wenatchee River, up again to Swauk Pass, where I had seen that lone traveler long ago. But then the route got blurry. I couldn't see where it went next.

Thea and I rejoined at the far end of the flat, where hobos occasionally camped beneath cottonwoods. As we wandered back toward the car, barn swallows hunted insects over the sward, western kingbirds and meadowlarks screeched and fluted from the few vantages. For practice I stalked and netted Oregon swallowtails, taking care not to damage their long legs and tongues. Even the fresh ones often lacked one or both tails, having lost them to would-be insectivorous birds. When a kingbird targets the waving tails with their orange and blue highlights, rather than the boring-looking body, the tails are doing their job for the butterfly's survival.

The place was full of butterflies. Little, bright orange mylitta crescents patrolled the pathways, darting out at purplish coppers the color of the vetch when the sun caught their prismatic scales just right. In truck ruts, where a little moisture remained, tawny skippers, buttery sulphurs, and cocoa wood nymphs puddled for salts. On the riverbank, a pair of Mormons were basking in the sun. Actually, they were Mormon metalmarks, very handsome rust, black, and white-checked butterflies, fall fliers and the only member of a vast group of tropical beauties that ranges up to the cold northern deserts.

We saw no more monarchs until, driving north on 97, a few flashed by headed the other way. They seemed to be moving with the river, all right. Soon I would be following them south, but first I wanted to reach their northernmost breeding grounds in British Columbia. It was early enough, I hoped, that most of the migratory generation would still be caterpillars or chrysalides. They would metamorphose into adult butterflies and begin their migration in September. We were seeing the advance guard.

At Pateros, where the wild Methow River gushes into the Columbia, we stopped at a fruit stand to buy late apricots and cherry cidersicles and to cast an eye over the milkweed around the mouth of the Methow. This stretch of the Columbia was once a major center for

Northwest butterfly studies. Andy Anderson in Pateros and John Hopfinger in Brewster maintained important collections and traded specimens with lepidopterists from all over. They explored a great deal of unsampled territory in the North Cascades and Okanogan Highlands, places like Twisp Pass and the Sawtooth Range above Lake Chelan, and were the first to collect several far northern species south of the Canadian border. Specimens with their Brewster, Black Canyon, and Alta Lake locality labels found their way into museums all over the world.

Hopfinger's and Anderson's records over more than half a century provide a window on the changes nature endured as the Columbia was altered. For example, in the early 1900s, Hopfinger found viceroy butterflies common all the way north to Brewster and beyond. Later, the inundation of riparian habitats by the Columbia dams drowned the viceroys' host plants and drove these butterflies up into the orchards. They were happy to feed on apple trees, but the heavy use of pesticides on the trees (little abated today) killed them off. Now viceroys can't be found nearly this far up the Columbia. Hopfinger and Anderson left no records for monarchs, however. It may be that milkweed, which spreads under some forms of human disturbance, has expanded upriver — and with it, breeding monarchs — even as the viceroys have withdrawn.

Showy milkweed is common around Brewster now. Late in the day we walked the shoreline of Casimer Bar to the confluence of the Okanogan and Columbia rivers. Young Mexican men, fishing with poleless lines, looked askance at Marsha. Riparian plantings of alders and osiers, peachleaf and curly willows — a floristic mélange of local species and imported cultivars vaguely related to indigenous types — show the restoration engineers' liberal definition of "native plants." Bright coppers shimmered in the pink knotweed, out of the wind. My broad canvas hat flew off into the marsh and I retrieved it with Marsha, another thing she's good for.

We camped at Osoyoos Lake State Park next to Vladimir and Vera, Czech immigrants from Vancouver. I was struck by their given names, the same as those of the great novelist/lepidopterist Vladimir

Nabokov and his wife. Nabokov, in his memoir *Speak, Memory,* expressed his amazement at how few people notice butterflies. My experience supports this and further suggests that of those who do, many refer to all large, colorful butterflies as "monarchs." Tiger swallowtails, in particular, are conflated with monarchs by the general public. Given the distinctive pattern of monarchs, this never ceases to surprise me. It speaks of people's vague awareness that big bright butterflies exist, the widespread familiarity of the name "monarch," and a desire to name the objects of the world even when knowledge lags way behind experience.

Vlad and Vera eagerly told of a "carpet" of monarchs on the ground at a rest area south of Nelson in the Canadian Rockies, but I knew that Jon Shepard, a lepidopterist in Nelson who knows the British Columbian butterflies better than anyone else, had seen no monarchs there. This was the first of dozens of intriguing but certainly erroneous tips I would receive in the coming weeks. Vlad and Vera had seen a mess of swallowtails at mud, or tortoiseshells, or fritillaries. Still, they had noticed.

As we curled into our sleeping bags, I thought again of the trip's first monarch, very fresh and richly colored, like burnt cinnamon and oranges. Easy to see at a distance, nectaring avidly on the purple vetch. I could have watched it for a long time, maybe followed it downriver until it was ready to fly on — though it might have stayed at Gallagher Flat all day or all week, tanking up for its long flight. I was sorry to drive it from the nectar fields. Such a fresh butterfly, so late in the summer, surely was a migrant. Any monarch emerging that late is subject to the dictate of day length to depart rather than mate. Wielding Marsha to no good effect, I might have sent it prematurely on its way to California or wherever it was bound; but it was probably ready to go. I wondered where and how it was spending this gusty last night of August.

Did an Irishman really put the "O" on what had been Soyoos Lake, as the state park interpretive sign said? The name comes from an Indian word, *sooyos,* meaning narrows. But the sign also referred to the painted turtles in the lake as "amphibians," so I wasn't sure I wanted to trust it on the Irish issue. Another sign told of fecal coliform present in the park water system, so we drank some of the last of our good Gray's River water. Vlad and Vera chose to ignore the notice, taking the waters of Osoyoos liberally. As we packed up and parted, we wished them well, and meant it more than idly.

Below the Canadian border, the Similkameen River flows from the west through a striking canyon. We followed it from Oroville to Shanker's Bend, then walked a rough track back along a canal that paralleled the river. An old homestead was now the headquarters of four species of woodpeckers: red-naped sapsuckers and little downies worked the bank willows, while red-shafted flickers and a family of blue-blacked, rose-breasted Lewis's woodpeckers all hung around the old hardwoods. Downstream stood an antique dam with a waterfall, where men and kids fished from the rocks. Steel-hooped wooden pipes the old watermen call "galleries" rusted and rotted their way down the canyon. Across the river a trail, built on the old Union Pacific line, ran upstream. Lined with milkweed and goldenrod, the trail rounded the rugged canyon and dived into a dramatic tunnel. On our side, a fresh Lorquin's admiral dried its big, banded, apricot-tipped wings on a sagebrush. This close relative of the viceroy had come out just that morning.

The tunnel emerged around a bend, pointing toward Nighthawk. Unwelcoming Nighthawk. From the cold stares and diffidence I have encountered there, I have sensed this tiny community with the beautiful name and remote location to be one of the least friendly in Washington. True, my first visit, several years earlier, may have been colored by the hordes of mosquitoes (suitably attended by nighthawks on the summer dusk). But this time the feeling was again palpable. No smiles, people disappearing around corners as we approached, no "Open" signs on the few businesses, and plenty of "No Trespassing" postings. Thea and I both felt it. Who can blame them, we agreed, with "Chopacka Estates" bulldozed out downstream and the thin edge of the cappuccino culture just a canyon away? Solitude like this is reluctantly surrendered. Garlands of wild hops draped themselves over ghost shacks with faded false fronts and sloping wooden stoops. This place too will change. I wanted to shout, Don't let it! The looks we got said, Just go away.

Beyond Nighthawk the Similkameen broadens into Champney's Slough, a marshy bottomland with oxbows. Two-tailed tiger swallowtails and coronis fritillaries thronged the abundant asters. There was *lots* of milkweed, but no monarchs; I was beginning to realize that large orange migrants were not going to be omnipresent. Well, I told Thea, a surfeit at the outset has spoiled many an expedition, though I couldn't think of any examples offhand. She replied that there seemed little danger of that.

The Similkameen River pours out of Canada. Rather than take the main road north from Nighthawk, I wanted to loop with the river and try a little boundary crossing that one of our maps showed farther west. The dirt road became a pair of ruts and stopped altogether at a fence well before the border. There was no entry station in sight. We lunched beneath Chopacka Mountain with a sheepdog who wandered by. The border collie (as, quite suitably, she was) begged bites. Together we watched a cheeky coyote trot over her field and across the international boundary, oblivious.

Looking up at the gray mass of Chopacka, the easternmost massif of the Okanogan Highlands, I recalled the day a few summers before when I had climbed it. The trail rose stiffly from lodgepole pines into heather and finally to a stony dome of a summit. And there, on the very peak, I found Bean's arctic: a wisp of wing a little darker than the wind. That lichen-and-granite-matching butterfly of the satyr subfamily was one of the ones that John Hopfinger and Andy Anderson first collected in the States. As far from being a migrant as a butterfly gets, *Oeneis melissa beani* clings to the cirques and arêtes of the North Cascades in tight little colonies, going nowhere until glaciers or global warming push them off. Such conservatism, showing ultrafidelity to habitat and place, is the exact opposite of the strategy employed by monarchs. The one stays put, adapting itself to a harsh but relatively consistent regime; the other comes and goes, chasing the seasonal conditions that make its life possible.

Whites coated the weedy mustards coming down from the border, and mourning cloaks, and the big Carolina grasshoppers whose black-and-yellow wings seem to mimic them leapt up from the dust as we drove past. At the shoreline village of Palmer Lake, a gorgeous garden beckoned, full of petunias, zinnias, and many other bright blooms, and spinning with colorful butterflies. I braked and hopped out. They turned out to be Leto fritillaries, the biggest orange butterflies in the region except for monarchs, and both Oregon and two-tailed swallowtails. All three species were swirling together in a fantastic tarantella, with an Oregon chasing a doubletail and the silverspot joining in.

Folks were friendlier here than in Nighthawk, and why not? The people had found what they'd come for, retirement in a community of their own making. As the agents of change, they were happy with the outcome instead of ruing the future and the certainty of unwelcome change. The gardener's husband told me to feel free to look around and, true to form, referred to the whole show as monarchs. Farther south, in Loomis, the proprietor of the closed gas station cheerfully came straight from his television football game to give us a badly needed tankful. But when we returned to the border, it was back to scowls.

The border crossing known as Nighthawk-Chopacka is a postcard scene, a pretty white cabin along a curving road into a broad vale. But the Crown Customs and Immigration officer, not particularly forthcoming — brusque, even — did not reinforce the bucolic hospitality of the place. "We'll dispose of these," he said, taking our Chelan apples and placing them well within his reach. He knew nothing of border-crossing butterflies and seemed almost affronted by the idea that they would do so without his permission. We didn't bring up the coyote. He finally waved us through, with a lingering look at our butterfly nets.

Heading up the Canadian Similkameen, we kept an eye out for fruit stands to replenish our apple supply. At Mariposa Organic Garden — *mariposa* means butterfly in Spanish — the proprietor, Lee, noticed my monarch T-shirt. She told me of a local family, the Mennells, who had something to do with monarchs. I watched an Oregon swallowtail dodging water drops on the lawn while Thea selected a melon, as the Mariposa was fresh out of apples.

Down the road in Cawston, we called on the Mennells. Robert, bearded and not far from forty, and lively, brunette Jane, both in jeans and work shirts, welcomed us warmly. Like most of their neighbors, they farmed organically, growing apples, pears, apricots, peaches, nectarines, and melons. They had just suffered a near-injury accident when six bins of pears fell to the ground from a truck. Even so, they insisted we stay.

They served us a fine dinner — their peppers and corn, our homegrown beans (the border guard didn't get *them*) — and fourteen-year-old Nicholas showed us his butterfly collection. He had done a splendid job of spreading them (better than I ever have, let alone at his age). I tried to impress upon him the need for punctilious labeling, a tedious business that raises a butterfly from a mere curio to a specimen of scientific value. For example, he had a beautiful white admiral — an eastern boreal butterfly of uncommon occurrence in western Canada, related to the viceroy and the Lorquin's admiral we'd seen that morning. Knowing it was from Conkle Lake, British Columbia, and the date of capture would make it a significant find; unlabeled, it could be from anywhere. I recalled trusting my memory at fourteen, too. I even remembered trusting my memory at forty.

A little pink-and-gray kitten rolled around and batted from within a coiled map, as old cat Pansy came up to me for rubs. My own pussycat, Bokis, was not long gone after nineteen years, and I welcomed the attention. Jane showed us a photo of Nick's nine-year-old sister, blond Claire, with a big smile and great ball of banana slugs in her hands — "She loves 'em," Jane said. Robert screened a long video of a monarch the family reared from a larva Nick had found in a nearby patch of milkweed. They had captured some excellent sequences of metamorphosis, as it grew and molted and transmuted from stripy caterpillar to jade chrysalis to flying butterfly. The release of the risen insect was clearly a major family occasion. I went to bed wishing that more young kids could be exposed to this kind of biophilia in their own homes.

On September 2, Robert Mennell rose at dawn to pick nectarines at 45 degrees F. Later he would salvage the bruised pears, to sell at a loss for baby food. The rest of us rose somewhat later. After breakfast we crossed an orchard to a cabin near a fruit warehouse to visit Ron Loiseau, a Québecois seasonal worker for the Mennells and an amateur student of butterflies. His collection was beautifully maintained, and I saw where Nick had learned his good spreading technique. Ron, lanky and black-mustached, showed us a big female Leto fritillary. Our northwestern subspecies is sexually dimorphic to a dramatic degree, the female a large cheesecake-and-chocolate confection quite different from the fiery orange male and unlike the tawny female of the related great-spangled frits that Ron knew in Quebec. He thought she might be a Nokomis, a prized species of southwestern desert seeps, and I was sorry to disabuse him of the notion. Ron said he knew a good spot for monarchs and offered to guide us to it.

Soon we were wading, dodging poison ivy, and kicking through knapweed on the Similkameen River dike beyond Cawston. The day was hot and humid, and my skin began to prickle. Nick and his younger brother, Chris, noticed feeding damage on milkweed leaves, and the little black barrels of caterpillar scat known as frass. Then he spotted a monarch larva browsing the edge of a floppy leaf. It was in its fifth and last instar, or molting stage, and thus not far from pupating. Spreading out along the shingle and the dike, we all began seeing monarchs at once, but we couldn't get to them — they were too far off, flying too fast, or across barriers of water or impenetrable vegetation. For the next hour, along the riverbank and through patches of milkweed, we ranged in the company of elusive monarchs.

Thea and I were walking along the dike in the shade of tall cottonwoods, when Jane called from a field off to the east: "Here's one!" I spotted the orange flash, on again, off again, as it crossed the field and settled into the lush foliage. Scuttling to the site while trying not to scare it, I spied the monarch as it rose: a russet vision like a fox darting from its green covert. I took an easy wing shot and netted the great butterfly over fresh milkweed bloom.

The monarch struggling in Marsha's diaphanous folds was a new female. Autumn emigrants are sexually immature and as a rule neither mate nor lay eggs until increasing day length the following spring triggers their hormones toward gonad development and mating. So she had only nectaring on her minute mind, and this she had been doing with fervor just before I caught her. Probably she had emerged only recently, perhaps maturing in this very milkweed patch or one nearby. But the milkweed's attraction for monarchs is not solely as the required caterpillar host plant. Showy milkweed, *Asclepias speciosa*, the only species in British Columbia, has thickly fragrant heads of florets like clusters of pale pink stars. Although happy to visit many kinds of flowers, monarchs take the nectar of their own host plants as avidly as any other.

The others converged to see the beauty up close, as I withdrew her from the netting with flat-bladed stamp tongs. Then I tagged her so that she might tell us something more, later in her journey, if we were very lucky.

Optimistically, I had brought along hundreds of tags in hopes of enlarging the sketchy western picture. My quarter-by-half-inch self-adhesive tags, like tiny mailing labels, came from the Northwest monarch study group known as One Thousand to One, based in Salem, Oregon. The one I now applied read like this:

MAIL TO
ENTOMOLOGY
NAT HIST
MUSEUM
LOS ANGELES
CA 90007
MP-2
81726

The final line is a serial number unique to each tag and thus to the butterfly tagged. When a recovery is made, just as with a banded bird, the tag establishes the data point showing that Monarch A traveled from Point B to Point C. Affixing the tag is a fairly simple procedure, but it takes practice and a firm hand. If you're afraid of hurting the butterfly, you may end up doing so or letting it get away. My personal technique involves holding the butterfly gently with my left hand, with the wings in closed position. I then grasp all four wings tightly near the base, with one forewing pulled up, the other pushed down. This is an easy position for the insect and does not hurt it. The wings are non-innervated chitinous material that can be handled gently without causing harm. Then, with thumb and forefinger, I rub the closed orange oval, or cell, bounded by black veins near the leading edge of one forewing to remove its scales. I used the left forewing, since most taggers use the right forewing and I wanted "my" butterflies easily distinguished from those tagged by others, especially if I should see them again.

Now, common mother lore has it that you can kill butterflies, or render them unable to fly, by rubbing the "powder" off their wings. But the powder, the shinglelike scales that impart the colors, are shed throughout a butterfly's life. Removing scales causes no real physical harm, although a fully scaleless butterfly would not be able to advertise its identity or pattern defenses to other butterflies or predators, and might lack some ability to shed rain and slip out of spiderwebs. But I remove just one quarter of a square inch of scales on both the upper and under sides of the wing so that the label will adhere to the membrane on both sides. I tug the tag free from its backing sheet, bend it over the stiff leading edge of the wing (the costa), taking care not to crimp the wing, catch an antenna, or stick two wings together, and clamp it down. I record the sex, condition (this was a fresh female with a bird strike out of the right hindwing), and tag number.

I released monarch #81726 onto laughing Claire's small nose. The butterfly paused just a second or two before her shining hazel-blue eyes, long enough for Claire to feel the tickle of the tarsi, before launching. Then the great insect rose, flew strongly off, and sailed over tall cottonwoods to the south.

The Mennells told me that in 1992 there were hundreds or thousands of monarchs in their valley, with larvae on every milkweed. After that they seemed to collapse here, as in fact they had throughout the West. The summer of 1992 had also been the last time painted ladies invaded the North by the millions, only to die off in the fall. The monarchs appeared with the flood of painted ladies, even well into milkweedless western Washington, almost as if milkweed butterflies had been caught up in the massive parade of thistle butterflies. Painted ladies burst out of the Southwest and Mexico every few years, when the spring nectar runs heavy and weather conditions are propitious. In the spring of 1992, Interstate 5 was closed in southern California due to the immense influx. "Good" years for both migrants do not necessarily coincide as they did in 1992. But I can well imagine that roving monarchs might be virtually bowled along by the sheer numbers of smaller orange-and-black butterflies in such years. Now, four years later, neither species was abundant. But there in the broad meadows of milkweed and tall grass behind Rocking Chair Ranch, monarchs were on the wing in British Columbia.

We took the larva back so Nick could rear, tag, and release it in ten days or a fortnight, along with any others they could catch. As we walked back, Ron spotted a huge Polyphemus moth larva, a fat apple-green serpent with silver embedded in bristled red spots along the spiracles, feeding on a willow. Then he pounced on a third-instar caterpillar of a western tiger swallowtail on a cottonwood sapling, its bogus snake-eyes pink and blue. Finally, he caught a big sesiid clear-wing moth *very* much like the yellowjacket queens it mimics. Northern crescents, like mini-monarchs, and big brown wood nymphs, like slow sparrows, flap-glided and flip-flopped through the waist-high grass. Nick and Chris caught leeches in a backwater pond, ignoring the mosquitoes. Like all country kids, they had their own taxonomy for the critters they knew: mozzies, river frogs, red flickers (the red-winged grasshoppers that faked us out again and again).

In the afternoon Thea and I pulled into Keremeos for supplies. At a rest area we watched a golden-tan satyr anglewing sipping at a urine patch in search of essential salts. It has made a completely different adaptation for surviving the Canadian winter. The monarch is probably the only species of butterfly that annually migrates into Canada in the spring and out again in the fall. Most butterflies get by in the egg, larval, or pupal stage, depending upon species. But the satyr anglewing, like half a dozen close relatives of the genera *Polygonia* and *Nymphalis*, hibernates as an adult butterfly through the harshest winters in a shed, bank, or tree cavity. Like the monarch, this one was tanking up on nutrients, but it was going nowhere. Heading upstream, we tanked up on the Mennells' nectarines.

The broad pastures and milkweed valleys narrowed to a piny canyon. Just before the walls closed in, Thea spotted a monarch pendant from alfalfa in a field with milkweed beyond Stemwinder Provincial Park. The crazy traffic on Highway 3 headed to Vancouver for Canadian Labor Day kept us from getting back to find it. This was the limit of milkweed, the head of the Similkameen's portion of the monarch migration, at least as far as breeding goes.

We took an early campsite on the south bank of the Similkameen, and slipped into the river's cool embrace under the looming presence of Bromley Rock. I held onto the shelf of a boulder and allowed the current to take me, like the sessile tadpole of a tailed frog. The polliwogs of that denizen of rapid mountain streams have suckers for holding fast to rocks so they don't drift away, and the adult males have a penis sheath in order to safely pass sperm in the current. As the sun dropped, the river grew still colder, and my own genitalia withdrew to about the stature of the tailed frog's equipment. My feet grew blue like the canyon's shadows. Then a mountain canyon campfire, the kind that calls for weenies & beans & beer. In the red flickers of the ponderosa flame I imagined monarchs rising into the night, flying off like sparks in the dark.

Thea's forty-ninth birthday took us back down the Similkameen and up its tributary, the Ashnola River, checking milkweed. The Mennells had seen several monarchs up in the high lake country above Ashnola. Well beyond milkweed, they were likely summer arrivals from the south. Immigrants will fly until they die searching for mates and milkweed and thus often end up outside their potential breeding range.

An impressively long, three-section, bright red covered bridge spanned the Similkameen near its confluence with the Ashnola. We live by the only remaining covered bridge in Washington, and we wanted to take a look at this one. Covered bridges were designed to protect the major truss beams from the elements. This type had no roof, but the side trusses were covered with siding inside and out, sandwich-style. Originally built for a railroad, it now carries cars. The Red Bridge, as it is named, is the northernmost location for the Mormon metalmark. But there were no butterflies at all about now, under the cool cloud cover.

The history of monarchs in British Columbia is obscure, and some suggest that the migration may be a recent phenomenon there. That might explain the absence of monarchs from John Hopfinger's collection earlier in the century. But at the Keremeos Grist Mill, we found evidence that their host plants have been there for many years. The historic house near the restored mill had an exhibition of photographs and botanical illustrations by Julia Rachel Price Bullock-Webster, an Englishwoman who lived there around the turn of the century. Julia's diaries had three references to swallowtails, probably the Oregon, which closely resembles the British swallowtail she would have known, and she noted "red, brown, orange ones" visiting zinnias. She made no clear allusions to monarchs, and her delicate and deft botanical illustrations did not include milkweed. However, Thea spied milkweed in one of Mrs. Bullock-Webster's photographs. A scene of Keremeos's Main Street in 1910, it showed a foreground of solid *Asclepias speciosa,* with leaves and pods clearly visible. So we know that monarchs were not kept out of western Canada by an absence of host plants in those days.

Later we explored stands of another plant that was long thought to be an alternative host for monarchs. We had crossed the watershed of the Similkameen, over Keremeos Creek, to the drier basin of interior British Columbia, and come to a place called Summerland. Neither sun nor summer was apparent. The mouth of Trout Creek was surrounded by brush, mostly Indian hemp, a waist- to chest-level, canelike herb with maroon stems and lemon-drop leaves in the fall. *Apocynum sibericum* (=*cannabinum*) is a type of dogbane used by many Native American groups for its fiber. The family Apocynaceae is closely related to the Asclepiadaceae (milkweeds); some dogbanes even produce a milky latex. Though Indian hemp has long been suspected to be a monarch food plant, recent studies by Susan Borkin in Wisconsin have shown it is not. Females may oviposit on it rarely, but larvae will starve rather than eat it. A single larva on Indian hemp in Utah is the only one I've heard about firsthand. Even so, the plant looked so much like milkweed that I couldn't help searching for monarchs. I didn't find any, but I did spot a brilliant picture-wing fly with a gemmed thorax; a striking yellow micro-moth; and a spotted prepupa in a shroudlike cocoon. There will always be interesting herbivores on any plant, if you look closely enough. "Brush" may be tedious when you're bushwhacking, but it is never boring.

The gray day went cold, very autumnal, and a storm came up with thunder and heavy rain. Instead of camping, we went to ground in the Lakeshore Memories Bed and Breakfast, and instead of more hot dogs, we splurged on fresh halibut for Thea's birthday. The rain bucketed down that night. But we were warm and dry, and, unlike Fred Urquhart's first versions in the 1930s, the new monarch wing tags are waterproof.

TWO

*O*KANOGAN

THE *OKANAGAN* VALLEY cuts through the center of British Columbia in a series of long lakes and finally the river that crosses into Washington, where the spelling and pronunciation change to Oka*no*gan. The name came from the Indians who gathered and potlatched at the river mouth. A friend, Helen Knight, who lives far up the valley in Clear Lake, had told me that in good migration years the monarchs breed all the way north to Kamloops, 120 miles north of the border, near the northern limits of milkweed on the western side of the continent. This year she'd seen a few adults, but no immatures, in the milkweed stands along the Clearwater. I decided to explore in that direction myself.

In Canada, you buy your gas in litres and get your change in natural history tokens — beavers on the nickels, loons on the quarters, robins on the dollars, kingfishers on the fivers. But money is still money, as the sprawl showed at Kelowna, in the waist of Lake Okanagan. Driving north, we passed signs for the "Better Than Nature" indoor gardening supply store; "Okanagan's Largest Park" — a shopping mall; and "The Natural Way to Live," which turned out to be a subdivision called Quail Ridge Estates. A billboard city sprouted by Bitterroot Bay where the pale pink flowers of *Lewisia* might once have bloomed.

In Kelowna, Butterfly World lay just beyond the paintball place. Butterfly houses, enclosed hothouses like conservatories or aviaries but for (usually exotic) butterflies, have sprung up all over in recent years, and of course I always stop for them when I can. Stepping into its hot breath, I dodged a great black swallowtail swooping down from the glass sky. The tropical air felt good after the cool off-lake breeze. It wasn't a large place, as butterfly houses go, but it was pleasant. Paths wound around streams, waterfalls, and islands of tropical vegetation such as banana palms and passion vines. The foliage was dense enough to carry off the ruse of a jungle glade, yet let in plenty of light to stir the captive butterflies into semi-free flight.

I was glad to see eggs, larvae, pupae, and chewed-up leaves right out in the open. In American butterfly houses, the U.S. Department of Agriculture, out of concern that alien species might get loose, requires the early stages to be sequestered behind glass and any eggs laid in the open to be quarantined. Although there are some grounds for these fears, especially in the South with citrus-feeding swallowtails, the rules don't make a lot of sense in northern areas where tropical species could not survive outside the hothouse.

Enormous blue-brown and eye-spotted owl butterflies (*Caligo*) and cherry-red *Cethosia*s fed on sliced oranges, and I got nose to nose with a *Caligo* drinking orange juice. You can't easily do that in Costa Rica, where they slip obsequiously in and out of shadows and outbuildings and land upside down on tree trunks. Their caterpillars, little straw wormettes with black heads, were lined up by the dozens beneath the banana leaves they had riddled. There weren't as many species here as in some butterfly centers, but lots of them were lovely longwings (red, orange, yellow, and black *Heliconius* species) flitting like bright lozenges above the ponds and waterfalls. Taped harp music accompanied two species of longwings playing tag around my feet. Pink bougainvillea hung overhead for nectar, luring big black-and-scarlet Asian swallowtails. *Catonephele numila,* a Halloween-hued buckeye, sucked on an overripe banana. A massive *Idea,* or rice-paper butterfly, alighted on Thea's head, where it seemed content.

Judy, the manager, released a freshly eclosed *Papilio ulysses* from a hatching cage, and watching that broad, unbelievably bright blue Ulysses swallowtail on its maiden flight into the low viny sky brought back to me the magical days I had spent in Papua New Guinea. Like the birds of paradise, the giant birdwing butterflies evolved there in genetic isolation. The constitution of Papua New Guinea calls for conserving these and the rest of the unique insect fauna. I went there to help develop a national butterfly farming program that would remove the pressure of commercial collecting while diverting the profits from expatriate dealers to the villagers themselves, who would then have a greater stake in keeping their forest lands intact as a source of breeding butterflies.

The rarest of the *Ornithoptera*, such as the dinnerplate-sized Queen Alexandra's and Goliath birdwings, were completely protected. When I had watched a female Alexandra, largest butterfly in the world, float high above me in the rain-forest canopy, I understood why she and the smaller, brilliantly chartreuse and turquoise male were so coveted by collectors. Their small broods and high value made them particularly vulnerable, and much of their habitat had already gone for oil-palm plantations and World War II airstrips that had grown back in coarse grass instead of rain forest. We concentrated our efforts on the commoner green and gold birdwings, hoping that the farming techniques might someday be well enough developed to apply to the largest, rarest species. The insect farming agency that developed in PNG has been emulated in many near-equatorial countries since. And while the specimen trade is still important, the largest market for butterfly farms is now for livestock to populate butterfly houses such as this one. A single chrysalis may bring several dollars, so every new butterfly house means more jobs in tropical lands — sustainable, resource-based jobs that help conserve human communities as well as wild habitats.

Visitors to the butterfly houses gain a sense of the wonder of these creatures and the need for their conservation. This Ulysses, for example, transported me back to the first one I saw glinting like blue Mylar in a flowering Highlands tree and every one after that, from a waterfall on Manus to the coral gardens of the Trobriand Islands. But as much as I like to be around butterflies, I would never want to run a butterfly house. It would be a lot like running a dairy farm with school tours every day at milking time. Before we left, I asked Judy if she'd noticed any monarchs in the valley this year. She said she hadn't seen one in years. "But then," she added, "I never get outside."

We stopped at a garden center called My Country Garden, on the southeast side of Kelowna, where scads of cabbage-white butterflies worked the rows of zinnias and marigolds. A purplish copper (color of an old maple-leaf penny on this dull day) visited Canada thistle in Canada. Stalks of tall droopy amaranth radiated electric Day-Glo cerise. I didn't see any monarchs at first, but a brilliant green, pollen-coated cuckoo wasp caught my eye on milkweed beside the road. Warmed by a shot of sun, it flew off, leaving a rain-spatter pollen patch stuck to the milkweed leaf. Looking at this yellow puff up close, I detected feeding damage on the leaf; then the striped design of a monarch caterpillar materialized before my eyes. That's how it happens: Now you don't see it, now you do. Right away I spotted another.

The first caterpillar's black tubercles, which people commonly mistake for antennae, were droopy. The second one had turgid tubercles that flicked up when I touched its back. I suspected the first one was going flaccid in preparation for molting. Lacewings and ladybird beetles foraged on the bright yellow milkweed aphids, making me wonder whether small *Danaus* larvae suffer some mortality on their account.

I was impressed to think of a female monarch finding her way to these traffic-blasted patches of milkweed, miles from any other, on Kelowna's outskirts, but I knew I should not be surprised. The ability of wandering monarchs to find their host plants over a vast, fragmented territory must never be underestimated. Again and again I would find immatures on small stands of *Asclepias* far from any others. Of course, it takes big milkweed meadows to support the bulk of the North American population; but as these outskirters show, no milkweed is safe within the theoretical range of the wanderer.

My hands felt and looked coated with Elmer's Glue after turning over many a milkweed leaf, and I wondered how milkweed feeders keep from being immured in the plant's acrid latex. Of course, it is the bitterness of milkweed milk that lends monarch larvae and adults their defensive unpalatability. Milkweeds contain varying amounts and types of cardiac glycosides — toxins that at low dosage may strengthen a defective heart's irregular beat but that paralyze the heart muscle in large doses. Emesis occurs at about half the lethal dose, according to the monarchs' master-investigator, Lincoln Brower. Monarch larvae and then the adults borrow the bitterness for their own use. The sap deters many potential herbivores from attacking milkweed. But monarchs and certain other insects have coevolved with *Asclepias* species to the point of specializing on them, taking in the cardiac glycosides without harm to themselves and becoming unpalatable in their own right. Brower concluded that the bad taste, vomiting, and effect on the heart all act in concert to discourage and condition predators to avoid monarchs on sight. When such insects are brightly colored to advertise their identity to birds that have tasted them once, they are known as aposematic, or warningly colored. The monarch is the classic textbook example of aposematism.

The proprietor of My Country Garden told me that the roadside verge had been mowed — thank goodness not sprayed — in July. So the milkweed there had grown up since then, and the larvae were offspring of late summer arrivals from somewhere else. She said she had seen no butterflies at all, and I did not tell her that there were dozens, of several species, out there on her flowers even as we spoke. Then she added, in a wistful echo of Judy at the butterfly house, "I scarcely get outside from May to November."

"Getting out" has always been a sine qua non for Thea and me, an important basis of our marriage. Now that we were facing a long period in which I would be out every day by myself, following the sometime chimera of the monarch butterfly, this Okanagan ramble took on extra significance. Over the next couple of days we went well afield even in the absence of monarchs. Uplake at Kalamalka Provincial Park, we searched among the pure scarlets and maroons of sumac, the soft oranges of wild cherries, for the animated scraps of autumn orange that meant *Danaus.* Above Lake Okanagan's south shore, white-winged crossbills and Clark's nutcrackers sought winter fat among the ripening cones of firs and pines. Up the steep slopes, the fragrance of yesterday's rain clung to the bunchgrass as we brushed past it toward stilling views of a quiet bay and the lagoon at its head. Still farther north, at Swan Lake, swallows massed to head south. The far shore, indistinct with the pallid morning sun behind it, reminded me of Lake Geneva below the Jura Alps — vineyards, orchards, villages, poplars. But in spring, instead of the wild crocuses of the Jura, there would be balsamroot, lupine, bitterroot. Magpies and starlings conversed musically in old willows. In marshes at the north end, bluets, those damselflies like little black matches with turquoise tips, hawked for gnats.

Veering south again, we entered Okanagan Indian Reserve No. 1 on an unposted dirt road. A flock of yellow-headed blackbirds, handsome birds seldom seen en masse, perched in willows. As we watched them, an Indian man with a long black braid approached. He confirmed that milkweed was sparse this far north. "I don't know any big patches," he said. Rounding the head of the lake, then stalking over Finstry Delta Park on the western shore, we saw nothing to alter that impression. Walking old farm lanes through abandoned orchards, cottonwood copses, and broad, tangled fields, we saw every weed known to man but milkweed. Although some of the comely native bunchgrasses swayed in the breeze, much of the old ranchland from which these provincial parks were carved was afflicted with a cover of alien plants: prickly purple knapweed, Saint Johnswort's rusty stain, tall black mullein stalks, Dalmatian toadflax. A handsome, unfamiliar shrub that stood out among the dominant black hawthorn, a barberry with clusters of fruits like little peaches, turned out too to be an escaped exotic. It's likely that we were carrying hitchhiking seeds as well. And the butterflies struggling against the cool onshore wind for warmth and nectar among the dry weeds were woodland skippers, clouded sulphurs, coppers, crescents, and cabbages — the lingua franca of summer's end in the West.

Monarchs did not reappear until we were again south of Kelowna and Penticton, skinning the eastern banks of Skaha Lake. Earlier Thea had spotted two incredibly cryptic caterpillars on a rabbitbrush: two inches long, with parallel lines of light and dark green and cream on a gray-green background, they were all but invisible against the ropy foliage. Since the cream, yellow, and black-banded caterpillars of monarchs proclaim their toxic presence instead of blending in like those camouflaged rabbitbrush browsers, they can be seen from a short distance in passing at some speed. Now, just after six o'clock, Thea cried, "Larva!" from the shotgun seat. I braked fast and we quickly found six larvae on one side of the road, two on the other, all in the fifth instar. They chomped impressively, carving ever-growing scallops out of the fuzzy leaves of the short, postmowing milkweed. I suspected they were all offspring of the same wayward mother.

A little downstream from the village of Okanagan Falls, we arrived at the home of Dennis St. John. Tall and slender, around fifty, with a silver-blond goatee and a thinning ponytail, Dennis looked a little like a stretched-out reflection of myself in a carnival mirror. As we settled in, he offered a good local beer and showered me with probing questions and trenchant opinions on a range of butterfly topics. "You know Nabokov's system of counting the rows of scales in blues?" he asked.

"Sure," I said. "He could tell the subspecies of the orange-bordered blues by fixing specific spots in relation to the rows of scales that radiated out from the base of the wing."

"Has anyone carried on that work?"

"Nah, others tended to think it was sort of like counting angels on the head of a pin: a little obsessive and pointless. They've missed its diagnostic value for showing and measuring locally evolved variation."

"It's helpful at a grosser level too," Dennis said. "I've found that of the two species of marblewings here, *ausonides* has its scales arranged in rows and *hyantis* doesn't."

The conversation went on through several beers and subjects. For Dennis, butterflies were partly a bright and beckoning escape from some difficult and sad realities. His adult daughter Elizabeth, of whom he was clearly very fond, was confined to the sofa because of a cruel and debilitating illness. Fortunately, she was a voracious reader. During our visit she joined in now and then, tossing out bright chips of conversation and wit. She came home to Dennis's house whenever she could, but later that evening he had to return her to the hospital. Thea and I spent the night in Elizabeth's room, among her books. As it happened, she would not live out the year. And when we saw Dennis the following fall, he was setting out in search of butterflies and distance from his grief.

But the next morning, Dennis enthusiastically took us to a site near Inkaneep, where a gardener friend of his had reported seeing monarchs. In an extensive field of milkweed we found only one big larva. The leaves of a nearby plant showed the crescentic feeding patterns of monarch larvae, but also revealed a big green *Mantis religiosa* with its grasping front legs extended. This exotic Chinese praying mantis, introduced by organic gardeners over much of the continent, is known to take adult monarchs in spite of their chemical defenses, sometimes rejecting the catch and other times gobbling it. Milkweeds vary in their toxicity, and monarchs thus display what is known as a palatability spectrum. Some monarch predators seem to be able to distinguish between the nastier-tasting and more tasty individuals. It isn't known whether mantids can do this or not, but I wondered whether they might have cleaned out some of the caterpillars here.

A dramatic rock gap yawned between MacIntyre Bluff on the east and Eagle Bluff on the west, cut by a much larger Okanagan River in former days of rapid glacial melt. Many of the monarchs flowing in and out of British Columbia must pass through this bottleneck, a major constriction for migrating birds as well. South of Oliver they would arrive at the Oxbows Ecological Reserve, where wetlands were being restored by ditching water back into the area. This drives old farmers crazy but delights birders and waterfowl hunters. But it was too cold for monarchs. Little white heath asters nodded in the cool gray breeze, unattended by butterflies. We crossed back into the United States at Osoyoos and for the rest of the day followed the Okanogan through Washington's orchardlands, where a much greater threat than mantids faced monarchs passing this way.

British studies have shown both lethal and sublethal impacts of insecticides upon small tortoiseshells and other butterflies. If they don't die, their breeding potential and vitality may be reduced. Orchardists use more chemicals for apples than for any other American crop, according to *Audubon* magazine. Thea, who once owned a small orchard downriver and later worked for a tree-fruit laboratory in Wenatchee, was well acquainted with the human concerns of life in a pleasant but pesticide-laden district. Certain cancer rates climb in apple country, and butterflies decline. Plenty of butterflies could be found at the Mariposa and Mennell organic orchards up the Similkameen, but the borders of the orchards we passed now seemed largely barren.

It isn't just farmers' poisons that insects have to deal with. Washington's Department of Transportation uses more herbicides to control roadside vegetation than any other state. Obviously, Okanogan County sprayed instead of mowing, too. Monarch watchers have noted that herbicides along the highways destroy large quantities of milkweed and nectar plants. Since butterflies commonly follow roadsides as corridors between larger habitats, the chemical barrier may affect the migration profoundly. In Britain, ecologists have supported a national roadside verge management program that seeks to prevent untimely mowing and deadly spraying. Careful verge maintenance has been credited with bringing back uncommon species that use these linear habitats, such as the British kestrel and the cowslip. No such enlightened ways were apparent in this valley. No wonder John Hopfinger saw the viceroys retreat southward.

The pesticide drift failed to reach every corner of the canyon. A milkweed patch some distance from any orchards produced a single larva, whose antennalike tubercles sticking out from under a leaf gave it away. Looking around reminded me that the monarch is only one of several common predators on *Asclepias* plants. A whole guild of distasteful insects occupy milkweed meadows across the continent. Two species of milkweed bugs, three species of milkweed longhorn beetles, several aphids, tiger moth larvae, and others share monarch habitat. The bugs, black and white and coral-red, like box-elder bugs, and the beetles, brilliant deep rose with black spots, also share the monarch's aposematism: warningly colored for birds' benefit, they too pack in the cardenolides and repel predators. It occurred to me that with all these bad-tasting insects on milkweed, you could consider the *Asclepias* herbivores one big Muellerian mimicry complex: a group of organisms that combine efforts to reinforce predators' reluctance to feed on them. A bird could learn to leave the whole plantful of colorful creatures alone.

I had hoped for a break from the orchards, hoped for broad bottomlands full of milkweed, but by five, I was about to despair of the day. Then, dropping down to the riparian floodplain of a wildlife management area, we found the most extensive stand of milkweed we could remember seeing. Ponds tucked among tangled vegetation rippled with the moves of coots, pied-billed grebes, gadwalls. I found several plants with the central top leaf stripped away but no larvae, then, feeding lower down on one of them, a woolly bear, or Isabella moth caterpillar. I knew that these popular black-and-russet hairballs, like most tiger moth larvae, were not picky, but I did not know they would feed on milkweed.

On the same plants I found some elegant silk-and-leaf houses made from one, two, or three leaves folded into a pyramid and laced in place with silk. I teased open a three-leafer to find a generous, sticky silk cocoon. The gap disclosed a mass of recently hatched spiderlings, their shiny short legs probing the sudden light. Feeling like a beast for disturbing them, I wrapped the package back up, interlaced the leaves, and glued them together with milkweed latex. I was thus occupied when I heard a high-pitched shriek, like a magpie's, from behind a cottonwood grove. Then I heard Thea call, "What's my prize? I've got a chrysalis!"

High up in a yard-tall milkweed, just four inches from recent feeding damage on the uppermost leaf, hung the pupa. It was pale jade, studded with brilliant gold flecks backed by black, and with fine dew droplets all over its lustrous surface. Suspended beside a ripe milkweed pod, it was wonderfully camouflaged against the backdrop. To the finder of the first chrysalis was to go a prize of the other's devising. I hadn't yet figured out what it would be, but I was beginning to realize how severely I would miss Thea's sharp eyes. Next she found a pair of plump caterpillars, a plant or two away from where they had been feeding. Often they spend the night on a new plant, perhaps to avoid drawing attention to themselves. Just a hundred feet or so from the pupa, they might have had the same mother, who might have eclosed here herself or spent a few days in passage from farther south. A sunny day would surely bring adults.

Thea's prize was dinner at a mobile taqueria in Tonasket. These chrome hot-food trucks turn up all over the fruit country, serving the heavily Hispanic migrant population. Maria's Unit No. 9 out of Bakersfield served up memorable tongue taquitos, downed with Okanagan Pale Ale smuggled in from the northernmost point of the journey. We camped for the night in sight of the few lights of Tonasket, high up in the national forest off Siwash Creek and the Havillah road. It was a shaggy forest, burned, cut, and thinned, grazed, ground, and cow-pied. But, like the wildlife management milkweed field, it was ours — public land — a relief after a day of "keep out" signs. A mass of stars unfurled overhead, promising sun and monarchs. But the day came with a cold overcast, heavy rain soughing into mist. Thea returned from a morning walk wet but with her hands full of fat shaggymanes, which, with coffee and a B.C. kipper, made breakfast.

Cruising the west-side road along the Okanogan River, we caught the cidery scent of ripening Red and Golden Delicious apples mixing with the wet, weedy smells of the canyon's intact riparian core. Across the river at Janice we crawled up a wisp of a road past Keystone. Mexican workers hung around a white clapboard pay shack. The roadlet, my atlas notwithstanding, did *not* go through to anywhere. But before it evaporated into an orchard, it gave a view of the canyon like a rough cut in aged and marbled meat.

Back out on the highway, campaign posters sprouted among the old asparagus and mullein stalks, one more sign of the season. The candidates included a Sump, a Roach, a Crapo, and a Paul Newman. Why did I imagine that Paul had better prospects?

Then down Wagonroad Coulee to Riverside, the head of navigation on the Okanogan River until the train came. The first Okanogan County Fair was held here in 1905. Today it was taking place in nearby Omak, but we skipped it. The county seat is notorious for the Omak Stampede, a bizarre event where cowboy and Indian horsemen and their unfortunate mounts plunge down a near-cliff of a course. Of greater interest to me was a handsome Juba skipper nectaring on a yellow chrysanthemum in a flowerpot by a gas pump at the Chevron. Its pearl-studded golden triangle looked as out of place as a nugget in a plastic pool, but it was not unusual as an example of butterflies' expert foraging abilities. Downcanyon, there awaited another such lesson.

Rattlesnake Point consisted of a shack and trailer compound with an amazing array of old cars, trucks, and tractors and a "Beware of Dog, Watch for Rattlesnakes" sign. Neither actually appeared, nor did any other living being. But in the next canyon, the rain-damp road and the warming day brought out many a butterfly, hoverfly, bee, and wasp. Davis Canyon is a Nature Conservancy preserve, acquired chiefly for its rare bunchgrass communities. In the canyon mouth I smashed a rotting peach on a rock, and it took less than one minute for the first lesser wood nymph to find it, along with the bees. Soon it would be thronged. I placed a female large wood nymph next to the lesser, and she went immediately from a passive mode (many butterflies play possum when caught) to avidly tonguing a vein of the sweet juice. Whether it is nectar for themselves or host plants to lay their eggs upon, butterflies are astonishingly adept at finding the right resources. You would think that the general tangle of plant matter would prove difficult for a mere insect to sort out. But first closing in by sight, then determining finer differences of scent or taste by scratching and palping the leaves with their tarsi and antennae, butterflies and other flying insects are remarkably good plant taxonomists. They have to be, since plants and insects have coevolved so closely that for many herbivorous insects only certain browse will do. And though lepidopterans are less constrained by nectar choice (many types serve well), they still need to locate plants with suitable corolla depth for their proboscis length, nectar abundance, and perching platforms. Not that these resources are always rare. On a state game tract on the river bench beyond Monse we first found yellow rabbitbrush bursting into generous, pungent bloom, set off by low purple asters. A persistently courting bumblebee owned one big rabbitbrush, orange sulphurs another. This prolific bloomer would paint vast sections of the roadsides over the rest of my journey.

Here too were the ancient remains of last spring's balsamroot, dried beige wheels of two or three hundred stem-spokes laid out flat, their splayed leaves in the middle. In spring, balsamroot colors the hills and flats an even brighter yellow than rabbitbrush, yet in my experience, very few of the numerous spring butterflies take advantage of this mass florescence. How a whole hillside of yellow composites can go wanting, yet a single ornamental chrysanthemum in a gas-station pot pulls in a thirsty skipper from the gods know where, is part of the mystery. There may be something about *Balsamorrhiza* nectar that butterflies just don't like. But as I would see again and again in the coming weeks, acres of available nectar might attract no wanderers, while a single isolated plant would have its bright flowerhead eclipsed by the welcome orange form of a visiting monarch, dropped from the sky as if in wait for my passing.

We crossed the Columbia below Bridgeport and swung back downstream to a point opposite its confluence with the Okanogan. Across the mouth of the Okanogan was Casimir Bar, where we had walked a week before. Searching for a way into the Wells Wildlife Area, we encountered a couple having a heart-to-heart about Jesus by the roadside. Practicing Christians, they actually offered us — complete strangers — the use of their canoe. I was tempted to accept, but between the late hour and the offshore whitecaps, decided to thank them and walk through weedy fields to the willowy shore. Common checkered skippers visited common bindweed, and woodland skippers nectared on cerise fireweed, another common flower I've almost never seen a butterfly use. A creamy marblewing larva grazed on a weedy tumble-mustard, a monarch larva on a milkweed overhanging the water's edge. This, the smallest larva yet, was nonetheless fifth instar. It was revealed to me backlit beneath a floppy leaf, betrayed by its late sun shadow against the diaphanous green tissue. This single caterpillar connected monarchs all along the Okanogan from the upper valley in Canada to the river mouth at the Columbia. It was also the first *Danaus plexippus* ever recorded for Douglas County, matching Thea's Chelan County record from Gallagher Flat.

The next morning, crickets and yellowjackets held forth on the flat. Thea's children, Tom and Dory, joined us from Chelan to swing their rusty nets, but the cool, dun day, with blue to the west, didn't call up the many butterflies of the previous visit. Thea spied a monarch larva the size of a baby link sausage in an aging, yellow colorfield of milkweed, and then another chrysalis, but this time her exclamation was more muted. It was damaged or diseased, leaking fluid, turning spotty. Back at the truck I dissected it: orange goop, very little structure. Monarchs, like other butterflies, host polyhedrosis viruses, bacteria, fly and wasp parasitoids that lay their eggs in or on immature stages, and a wide range of predators in spite of their unpalatability. I was surprised that this was the first afflicted individual we had found. A pinprick hole suggested that it had likely been attacked by a tiny braconid wasp that pierced the exoskeleton without actually succeeding in leaving eggs, causing it to leak hemolymph and rot.

Monarchs sustain a protozoan parasite, *Ophryocystis elektroschirra*, at sometimes devastating levels. Infected adults carry the spores, which they leave behind wherever they alight, even on their own eggs. Hatching larvae often consume their eggshells and thus take in the spores. When the caterpillar pupates, the parasites develop in the tissues that will become adult scales. While many monarchs are infected, the rate being heavier in the South than in the North, they can usually function normally with light infestation. Severe infection causes developmental failure and death. I burned the remains of this pupa so as not to spread any pathogen it might be carrying.

No more living, soaring monarchs graced this visit, only a loose flock of colorful ultralights sailing over from the Chelan airport up on the bluff. The motorized hang gliders looked graceful until I pictured our first monarch rising through the same airspace. In comparison with that cinnamon vision, the ultralights were clunky, inefficient, garish gliders. They might be useful for following geese, or leading them. But it would take a flying machine never yet brought off the line at Boeing or anywhere else to fly with monarchs. It would have to combine the traits of a helicopter, a Stealth jet, and a weather balloon with the grace of a swift and the weight of a single feather. The only craft suitable for following monarchs aloft would be another monarch.

Mine would be a groundborne pursuit. And, soon, a solitary one.

Thea and I would share just three more days afield before parting for a long time. I had stowed my car in an orchard belonging to family friends during our preliminary travels in Thea's truck. When we returned to fetch it, Martha, pregnant and due in two weeks, was gathering fallen apples with her daughter, Tirrell. We joined them, stocking up for the road. Nearby, a mourning cloak was sucking on sweet Galas, already rotting.

A fresh mourning cloak is a bright and lively thing, concocted of chocolate wings, French vanilla borders, and blueberry spatters near the margins. But when December seals these orchards in icy riverbreath, the hardy cloak will be snug in a shack or a cleft somewhere, its wings closed like an ashen cowl. Metabolism sunk in a long winter's nap, antifreeze in its tissues to protect against damage from ice crystals, it will keep its life in reserve for spring.

By the time the mourning cloak hibernates, every monarch in the land will have flown hundreds of miles to a different, milder winter. This we know. But where and how the monarchs of the Similkameen, the Okanogan, and the Columbia would go, I could only guess. We set off, in separate vehicles now, to let them show us.

THREE

Grand Coulee

In 1916, my grandmother, Grace Phelps Miller, traveled to her first teaching job in Chelan. Grammy had first taken the train from Seattle through the tunnel at Stevens Pass, then a riverboat up the Columbia from Wenatchee, and finally a stagecoach around Chelan Falls. The Great River of the West is even greater now (because dammed) but much less (being dammed) than when she steamed up its unimpeded reach.

The monarchs had given me no clear directions yet, other than occurring along the major south-flowing drainages. Since the Columbia no longer has steamboats and is largely impassable by land southward from Chelan, the promising way lay off to the east into the coulee country and thence to the pocked plateau known as the Potholes. From there I could rejoin the Columbia or carry on wherever the monarchs bid me go. Besides, a quick look-see up into Grand Coulee itself seemed prudent while I was in the area.

We crossed the river at Chelan Falls and drove up McNeil Canyon, over the blowing soils of the Waterville Plateau to U.S. 2 and Moses Coulee. Coulees are grooves in the basalt-flow flats, the channels in the channeled scablands, cut by catastrophic floods from breached glacial dams. Moses Coulee's little McCartney Creek, awash with weedy whites and sulphurs, gave no hint of the torrent that raged through it just thousands of years ago. We were headed for a bigger coulee, the biggest of them all.

A rabbitbrush flat up the east side of Moses Coulee looked promising for monarchs, but revealed instead a different orange-and-black toxic insect: many pairs of robust meloid beetles, mating most salaciously. These one- to two-inch, warty black beetles with acid-orange heads and thoraxes are highly distasteful and warningly colored, just like monarchs. I had first encountered their poison powers when I had several warts removed from my hands. One method was freezing with liquid nitrogen. But for the worst ones, the doctor applied a tincture that brought up the skin in vicious blisters, eventually spalling the wart. What he used was a cantharin derivative — an alkaloid refined from cantharid or meloid beetles, often known as blister beetles.

Another name for the beetles, and an extract made from them, is Spanish fly, the infamous "aphrodisiac." Spanish fly induces the same sort of extreme skin irritation as the wart remover, but it concentrates in the genital mucosa — hence the itch in the pants that poses as desire. But this nasty beetle juice can cause tissue damage and even death in large doses. Now, watching these fornicating beetles, I wondered if their frank behavior did not explain the folklore as much as the effect of the substance itself. Certainly their bitter flavor allows them to indulge in such orgiastic display in broad daylight, without a worry for lark or sparrow. The same principle applies to monarchs.

Cresting the plateau, we passed Sagebrush Flats, scene of a bitter land-use battle between ranchers and conservationists trying to protect the state-endangered pygmy rabbit. The only North American rabbit that digs its own burrows, this species depends upon mature sage steppe in the Northwest interior. Between the cow, the plow, and the Columbia Basin Irrigation Project, there is not much high-quality sageland left. The best pygmy rabbit site on public land was recommended for protection by the Natural Heritage Advisory Council, on which I sat. But predictably, in a district where it is said that every cow has two votes (and bunnies have none), the cattle interests won out when the state favored a private grazing lease over a natural area preserve.

A dramatic side coulee ran from the flats down to Bower's Spring, where seeps flowed against the rimrock. Once a rich oasis, it had been grazed flat and was now full of the weedy species of plants and butterflies familiar throughout cattle country. Nothing has had a greater negative impact on butterfly diversity in the West than cattle. Of course, there is such a thing as responsible grazing, and I have seen rangeland that supports a full range of native species as well as an appropriate number of beef cows. But this land, part of the holdings of the same rancher who got the state lease on the rabbit habitat, was hammered. Since milkweed thrives under moderate disturbance, monarchs can do somewhat better with livestock than many other butterflies, but there were none here. Milkweed was virtually nil; two plants struggled to compete in a dry streambed choked with coarse and stickery knapweed, tumbleweed, and pigweed.

Moses Coulee appeared again through the cornfields and feedlot of the rancher in question. Old tanks and farm machinery were lined up under a lava flow for a hundred yards or more: an archaeological lode for future diggers, like the stash of early Clovis spear points found not far from here under a basalt overhang. Which direction the culture has taken, between Clovis and John Deere, is an open question. The road was public, but a woman by the fence gave us a grimace reserved for trespassers and bunny-lovers.

Palisades was an isolated handful of houses, a green square with big, kempt shade trees, and a pretty brick school, all surrounded by columnar rimrock and rubble in angles of repose. The towering black cliffs were the result of the Grande Ronde lava flows originating far to the southwest some 16 million years ago; in some places they are a mile deep. Dark stripes of basalt ran among the various whites of snow buckwheat, clematis seed puffs, leached alkali, and the bird lime of hawks, owls, and ravens.

Moses emptied us onto a main thoroughfare through unseductive farming towns like Quincy, Ephrata, and Soap Lake. The weedy, sprayed, and burned roadsides did not look promising. When a lush stand of showy milkweed appeared in the lee of a ditch, I leapt out and quickly found one firm monarch larva, which dropped to the ground when I touched it. Many caterpillars go limp and fall, presumably as a means of escape. Among monarchs, I have found, some individuals drop and some do not. Variable behavior, like variable appearance, can be the result of what population biologists call a "balanced polymorphism." Another example would be the chrysalides of anise swallowtails, a certain proportion of which turn out leaf green while the rest are wood brown. It seems that natural selection supports a range of traits; the mix provides for a higher rate of survival than any single formula for all would impart. I placed the larva back on its leaf.

Grand Coulee is just a big dam in most people's minds, but the landform itself is much more. The Purcell Trench lobe of the continental glaciation flowed between the high walls of the Clark Fork in today's northern Idaho, forming a 2,000-foot-high ice dam that backed water to the east over 3,000 square miles — more than Lakes Erie and Ontario combined. Each time the lake neared the top of the dam, it floated the ice and poured out beneath. In this way, Glacial Lake Missoula emptied itself many times between 12,000 and 16,000 years ago. The resulting catastrophic floods sent thousand-foot walls of water at fifty miles per hour through the Columbia Basin, cutting the coulees like an ice cream scoop. The Okanogan Ice Lobe diverted the ancient Columbia through Moses Coulee and steered much of the floodwater through Grand Coulee.

As we entered the great trench of the Grand Coulee itself, loons bobbed on Blue Lake. They had bred much farther north. If this rugged slot was a flyway for loons, who else might be using its corridor? I suspected that Grand Coulee might be a second major thoroughfare for monarchs passing to and from Canada, paralleling that of the Okanogan to the west, across the generally inhospitable wheatland barrens of the Columbia Basin. Although I was generally southbound, I fell for the lure of reconnoitering up-map again briefly to check out Grand Coulee. But it was also a gamble to go north again, when the days were shortening. What if the monarchs left me behind?

Like a fire on the morning air, the monarch dropped from the sky, blazed against the lake, and settled on the cooler yellow flame of a common sunflower. I was shocked. I had never, as far as I could remember, seen any butterflies whatever nectaring on sunflower. But there it was, a very fresh female, definitely digging in to the flower head's nectar tubes one after another with her supple black drinking straw. Thea was nearest but, concerned she might miss it, she looked my way. I nodded. She had very much hoped to catch an adult before leaving, and she couldn't brick it any worse than I had at Gallagher. A deft stroke, and the monarch was safe within the soft nylon folds of her net.

We had driven up Lower Grand Coulee with its unflooded flats and natural shorelines, strung out along a series of lakes that lie in the creases left by the ancient ice-water scoop. At Sun Lakes State Park, hard little pears and yellow plums plopped to the ground beside our camp. Here Thea and I had once released and photographed on rabbitbrush a monarch I had reared from a caterpillar given to me by a child. Every other time I had been here, the lakes had been packed with noisy boaters and fishers. Now, after Labor Day, the campground was nearly empty. The few neighbors included a gritty pair with a metallic blue Harley trike and a black-hatted man who sat at a picnic table playing with a large knife for hours. Before dusk, something orange shot past on a wind that rocked and shook the car all night.

In the still dawning, a sharpie chased a flicker through camp, as a pair of mule deer cleaned up the plums and pears that had fallen since the raccoons' night harvest. Loons dove offshore, goldenrod and aster nodded onshore in a clear and sunny morning. We took a clinker trail up into the Roza member of the Yakima Formation, a Late Miocene basalt flow that spread over some 20,000 square miles at 25 to 30 miles per hour, covering it to a depth of 100 feet. Lenore Lake Indian Caves, sucked out of the Roza flows by whirlpools, were occupied at least 5,000 years ago. Today they offered but harsh comfort, their sharp cobble floors heating up under black-rock eaves painted with lime, chartreuse, and orange lichen and graffiti. Woody penstemon briars pruned by deer showed one last pale pink blossom. Two couples clambered up and snapped each other, all in blue T-shirts proclaiming "World Peace and Unity." When they left, I stood among the broken boulders and surveyed the shoreline below. I saw a band of nectiferous flowers, and dropped down for a look. That's when Thea, then I, spotted the monarch.

I quickly set about tagging and examining the very fresh female, perhaps eclosed that very morning. The burnt-sugar wings still carried the blue-flame iridescence that only the most innocent monarchs possess. Though it doesn't hurt the butterfly, I dislike rubbing even the small area required for the tag clear to the membrane. As Jean Craighead George wrote in *The Moon of the Monarchs,* the edge of the wings "seemed as if the night and day had been knitted into them; they sparkled white and black." The head, with its fine coil, its black globes, is just as brightly studded as the starry dark margins. When I was finished with the tagging, my fingers sparkled.

I felt the powerful flight muscles pulse against the vascular pads of my fingers, until she went limp in the possumlike passive resistance some butterflies affect. Then, stroking her prickly legs to get them moving, I replaced her on a fresh sunflower. She sat with her wings closed for a few moments as I backed away, then pumped them and launched southward some twenty yards to a sagebrush. There she basked, pumped, and rose again to sail south onto a past-bloomed sunflower, where she remained, opening and closing her wings, for several minutes.

Now we saw skippers nectaring on sunflowers, sulphurs too. I grew up among these same sunflowers *(Helianthus annuus)* in eastern Colorado, and never saw this. I did notice that nearly every one of the home sunflowers held its resident ambush bug. These cryptic, angular, predatory sucking bugs with mantislike grasping front legs frequently take butterflies. Maybe in Colorado the butterflies had adapted to the predators' abundance by avoiding sunflowers. Here they displayed no such rejection.

Local traits, at least in appearance, can and do evolve. The woodland skippers here in the coulee were remarkably pale compared to the chocolaty ones at home on the coast. Though only half the size of the Yuma skipper, they could be mistaken for that large, immaculately tawny species of the alkaline reed beds of the Great Basin, whose northernmost known colony is here at Sun Lakes. Locally evolved traits can also accrue invisibly at the levels of behavior and ecology. Monarchs look alike all over North America, and in the limited tests conducted so far, no differences have been detected between the DNA of eastern and western populations. Yet it is supposed that they observe radically different set destinations in their migrations.

This monarch on the sunflower, for example, looked not a whit different from a fresh female that at the same moment might be poised on goldenrod at Cape May or asters at Cape Hatteras. Yet it is widely believed that she relies upon some inner setting that will draw her ineluctably toward California, while the Jersey or Carolinian insect will carry on to Mexico. Merely an insect, it is said, the monarch is the pawn of parsimonious instinct and little more.

"I have tried, in my way, to be free," wrote Leonard Cohen, in a song that ran through my head. As I watched this perfect monarch steadying herself to sail on, facing a journey of maybe a thousand miles or more, I had to believe that there was a little more of something like freedom built into her system than the rigid lines drawn on migration maps seem to allow. Nothing so refined as freedom the way we think of it, when we imagine ourselves able to will where we go and what we do. Nothing as derived as "decision." But the freedom, nonetheless, to respond to the vagaries and gifts of the days with every adaptive tool and option that kindly evolution has provided.

The idea of the monarch as an infinitely elegant mechanism for transmitting its own germ plasm from one season to the next, no different from any other animal in that respect except for the baroque splendor of its means, does not alarm me. But to think that it has no latitude in its path, no play in its tether, is to cast this animal entirely too narrowly. As the lay of the land swells and falls, as the rivers turn and pool, as the winds unfurl, and as the trees and flowers grow, so goes the monarch. Merely an insect it is, but it is no simpler than the sum of all that its world offers. And, given essentially the same equipment, not all monarchs will behave the same way.

Monarch #81741 was caught at 11 A.M. At 11:30 she was still hanging there. She was not so obvious with her wings folded, but her white spots and her tag stood out brightly against the rising glare of the day, much like the luminant white checks on the necks of the loons rising and diving a few rods away. Monarchs must nectar prolifically to build up the lipid bodies that will both get them through their winter hiatus and fuel their breeding flight in the spring. Autumn migrants must therefore balance their imperative to get on with the trip against the equally pressing need to feed continually. I know the feeling.

The monarch moved again. Avoiding the heat, she alighted beneath the parasol of a broad, sandpapery sunflower leaf and drank from the last plant I expected to see her kind visit. Sweating and hungry, we watched as she worked the sunflowers exclusively. Careful not to drive her, I followed south a couple of hundred yards. She disappeared around a willow, and several red-winged blackbirds flew out — because of her, or me? I hoped there would be no strikes: I saw no need to train a young redwing at the cost of #81741. Working southwest along the shore, I lost her among the willows at 12:15. Finally, admitting she was gone, we left the lake, taking along the beautiful vision of that bright flier, flap-gliding against a backdrop of orange-daubed black basalt.

Clear water, grand implacable lava flows and rimrock, sharp black scree, mountain bluebird sky: the scene was much the same one the Indians had seen from their stony abodes above. Did they also see the monarchs moving through? That's hard to say. While we often speak of the migration as "ages old" or "ancient," we don't know for sure. British milkweed butterfly specialists Dick Vane-Wright and Phil Ackery, in their "Columbus Hypothesis," suggest that the great North American monarch migration may largely be a recent phenomenon, resulting from the postsettlement deforestation of much of the continent and the consequent proliferation of milkweed. But Lincoln Brower and others argue that the center of abundance of milkweed simply shifted eastward, that the Great Plains had more diverse and abundant milkweeds by far before the breaking of the sod. There is no conclusive evidence for either view.

In any case, the migration as we know it here has to be postglacial, arising sometime in the past 10,000 years, after the ice dams broke and cut the coulees. I can't help but wonder whether there were already people here when the great floods poured through. They would have been plucked from these caves like lizards in a flash flood, and any butterflies erased in an instant. But time restores. The antediluvian Lower Grand Coulee (unlike cattle-handled Moses Coulee and inundated Upper Grand) is today replete with milkweed and honey: a perfect migratory pathway for monarchs and me.

Lower Grand Coulee butts up against Dry Falls at its head. This dramatic rock wall, 400 feet high and four miles across, was worn back more than twenty-five miles by successive meltwater floods that gave it the greatest flow of any waterfall in the world. Migrants heading north up the coulee must top the huge basalt walls — no great trick for a butterfly known to fly in air currents thousands of feet high. Scaling those enormous cliffs, they would find a glen full of milkweed and sunflowers on top, next to Dry Falls Dam. The way along flooded Upper Grand Coulee winds from Coulee City to Electric City, combination old-time and dam-boom towns with more cinder block than brick, and past the monolith of Steamboat Island. Some 800 feet high, Steamboat is a "goat island" — a remnant of the lip of a cataract formed when the prehistoric current split around it and wore the rest away. Exploring for plants, botanist David Douglas passed the landmark in 1826, long after the great floods but before the dams that would make Steamboat almost an island again. Its flat top still supports rare plants that neither cattle nor reservoir has been able to reach.

As we neared Electric City, Grand Coulee Dam groaned in its broad notch off to the northeast. We, who know no other time, tend to think of the dam as the impressive thing around here. It is the biggest chunk of concrete anywhere, and it changed almost everything when it stoppered the river in 1936 — lives, land, fish, wildlife. Along with the rest of the dams that followed, both upstream and down, it made sure that life along the river would never be the same. Not so that you'd want to wait around for it, anyway. FDR's reward for building the dam, besides a hell of a lot of jobs, was Lake Roosevelt — the long impounded pool of what was the Columbia River. The price of hubris, the bill for attempting to mimic the glacial ice dams, is just coming due in the form of the Northwest salmon collapse.

The road descended through steep curves from the plateau to the water at Keller Ferry, about fifty miles east of the Okanogan River mouth. At the shore the Colville Indians run houseboat rentals and the National Park Service runs a campground. In the light of the bathhouse there, I saw several toadlets, less than an inch long, ambushing insects. The western toad *(Bufo boreas)* has been lost in recent years from many of its former habitats. Places where I used to see them commonly now reveal nary a toad. I remembered the title of a paper I'd heard at a recent conference on amphibians: "Another Nail in the Coffin of the Western Toad." It is widely known that many species of amphibians are dropping out of large parts of their accustomed ranges. Several factors are thought to be involved, including increased ultraviolet radiation from ozone loss, overexposing the skins and eggs of species that are UV sensitive; diminished water tables, causing reproductive failure; and toxic pollutants. Lately, a new amphibian virus has been proposed as a factor. So it was good to see these toads. An old saying: "Monarchs by morning, toads by night, such are the naturalists' keen delight."

Looking for that morning monarch the next day, Thea and I walked the beach below sagebrush and Chinese elms. As well as on the air, we watched out for monarchs in the beach wrack. I once saw windrows of drowned wanderers on the New Jersey shore when they were passing through Cape May in waves. Another time I fingered washed-up question-mark butterflies at Gay Head on Martha's Vineyard, following a southwester. Insects that move en masse along shorelines sometimes meet winds that smash them or waves that swamp them, and end up as such fine flotsam. Colville Indian kids busing south to school at Wilbur looked downstream at me as their ferry docked.

We boarded for the return trip, crossing the Columbia on the *Martha S.*, a little floating platform under power with metal treads down the middle and blue and white trim. There were just two other cars on board. Of the three remaining Columbia ferries, this was the middle one. We live near the lowermost, the *Wahkiakum* out of Puget

Island, and we would take the uppermost that afternoon. The far shore was brown rocks, beige screes, pale yellow slopes with black pine spatters and high hills beyond, all running steeply down to the water, five feet below its own high mark on the canyon wall. Blue sage and jays showed against a pale blue sky, blue elderberries hung in the dust.

Now we were standing on the Colville Indian Reservation, 12 million acres of montane habitat grudgingly set aside in 1872. I wanted to see the Sanpoil River, where monarchs were once recorded in the summertime. Beyond the few houses and stranded automobiles of Keller, blue willows and gray birch showed the course of the river. An osprey whistled around the bend. A golden eagle flew up and circled, and the much smaller osprey dove on it. Both shone richly in the sun. The delta of the Sanpoil was swallowed by the reservoir on the Columbia, leaving a muddy, cobbly beach. Patchy milkweed lined the ring-around-the-tub, drawn-down shoreline, but it looked unhealthy, most plants wilty and spotty with a rust, others lacy with predation. I wondered if such poor fodder would affect monarchs, like eating bad bread. There were none to be seen.

Water pipit and solitary sandpiper wandered the cobble; golden eagle from upshore circled overhead amid the separate cronkings of raven and great blue heron. On the point, at what you could call the mouth of the free-flowing river, stood a sort of shrine or marker, three slender pieces of driftwood bound together. Above it on the western shore, a yearling black bear scuttled scared along a track through the pines, looking over its shoulder with an expression something like guilt, or anxiety.

Small blue butterflies flickered over the dainty white asters upshore. The acmon blue bore silver-and-turquoise scintillae beneath, while the eastern tailed blue, as its name implies, sported tiny tails on its hindwings. That nearby pipit, were it to spot the blues and strike, would probably go for these bright diversions instead of the indispensable parts, just as with the much larger swallowtails. Tails and colorful bull's-eyes have evolved several times among butterflies as effective defenses. This kind of parallel evolution, even more common among

Lepidoptera than warning coloration, led to numerous presumptions of relationship that later proved false. The family of butterflies to which these little blues belong, the Lycaenidae (blues, coppers, and hairstreaks or, collectively, the gossamer wings), provides many examples. There are coppers and blues with tails, coppers and hairstreaks that are blue, and blues and hairstreaks colored copper. Yet their structure reveals several separate lines of descent, with minor traits such as color and tails added or dropped by natural selection like mere accessories for a wardrobe.

Thea noticed a Becker's white, a beautifully green-veined butterfly, darting at something purple along the shoreline. The color swatch proved to be the purple skirts of three swarthy, busty dancing maidens. Rescued from the mud and washed off, they were revealed as plastic wind-up Disney figures from Burger King, representing Esmeralda from *The Hunchback*. I could picture the Colville Indian child playing with the dolls on the shore as her family fished and accidentally leaving them behind; could hear her sobs. I saved them, and saved some future anthropologist from having to compare their cultural context with that of Clovis points. They joined the company of found objects adorning Powdermilk's dashboard, an ever-changing traveling museum of natural history.

The road across the reservation was a long, hot dusty washboard shared with close-passing log trucks. Many orange butterflies flashed in the heavy sun, but they were all silver-spotted; we followed coronis fritillaries, crazy for purple bee balm, all the way to Inchelium on the Columbia River. As we crossed back over on the *Columbia Princess*, a bigger little ferry with a pilot on a high bridge, a woodland skipper flew on board for the passage. I once saw a pine white butterfly do the same on a superferry in the San Juan Islands of Puget Sound—embarked at Shaw Island, disembarked at Lopez.

A modest number of monarchs reach southeastern England almost annually. Although prevailing winds, which also carry American warblers and sparrows across the Atlantic, seem sufficient to account for the crossing, some attribute the phenomenon to butterflies stowing away on transatlantic liners and freighters. Monarchs sometimes end up on oil derricks in the Gulf of Mexico. It is possible that mid-Atlantic waifs, having begun the trip on their own, make landfall on ships when they can, lending truth to both hypotheses. The skipper flew off when we landed on the eastern shore.

Lake Roosevelt, the former Columbia that elbows north now into Canada, was lined with yellow and red coreopsis, bright low daisies full of bright little butterflies. They marked without giving away the town of Daisy, a once-thriving community abandoned to the lake's rising waters. Briefly, we called on Rosie and Cecil Carr, folks I'd met on a previous trip because their big marigold beds had been full of butterflies. Now, after chicken dinner, they sent us off loaded with their own ripe cantaloupes and a vivid picture of their flooded hometown. "You can see the former streets and foundations when the reservoir is low," Cecil said. "It's an eerie experience if you used to live there."

Driving south, I could see that it wasn't only people who were displaced when all the flats and farms were inundated sixty years ago. It seemed clear that monarchs were driven from the upper Columbia as well. All the good milkweed habitat was drowned, the shore rising now right up to mud and pines and rocks. What does remain has largely been developed; Swede Flats, for example, has become Deer Meadows Estates and golf course. Though a few individual monarchs will continue to follow the great watercourse north to the border and beyond, they're not likely to have much issue. The Okanogan Valley probably accounts for most of the international traffic in monarchs both ways.

We camped at Fort Spokane beneath great yellowbelly ponderosa pines and stars, listening to the counterpoint strums of katydids and crickets. In the morning I swam to wakefulness where the Spokane and Columbia rivers meet. The previous fall, on a dry run, I'd come to this confluence and found not monarchs but another big rust-and-black emigrant, the rare Compton tortoiseshell. This overwintering nymph occurs all the way across the Canadian Shield from coast to coast, sometimes spilling out as far south as Oregon, other years pulling back. Its movements are inscrutable: no one knows why it fluctuates or how far individuals wander. Biologist Adrian Wenner, of the University of California at Santa Barbara, has proposed that monarchs similarly just expand and contract their range rather than truly migrating, that we have simply imagined an elegant pattern for them.

I didn't believe it. I felt that what Wenner called the romantic view, that monarchs engage in a true, largely birdlike migration, was reflected in the facts as we knew them. I also felt an irresistible compulsion to join that passage. Investigating "facts" was just a part of it. But for now, parting was upon us. We climbed the winding road out of the canyon toward Creston, parked to look back at the river, and kissed goodbye in a wild wind. Thea turned west as I drove south.

I knew that I had no choice. No matter what they shared with me or withheld, however many I saw or imagined, I would follow monarchs until I could follow no farther, or until snow sealed away the season for good.

FOUR

Crab

The way south was a very straight line across the sagelands and wheat barrens. Sinking Creek, Sinking Fast Ranch, a turn-of-the-century cemetery of withered lilacs and lichened stones, its centerpiece tomb entered by badgers. Partridges bursting from peppermint fields. Dust devils and grain elevators looming over all. When you fly over these flatlands, you see great circles tangent across the basin, like cookies cut from a golf green. On the ground, elevated pipes endlessly roll on wheels to sprinkle the circles, engines of the Columbia Basin Project, designed to turn deserts into gardens. But with their brown furrows, black stubble, and yellow straw to the edge of the road, these fallen farmlands are monotonous even to a biologist. Only horned larks seem happy here.

When the landscape offers no landmarks, no rivers to go by, and few patches of flowers to draw them on, how do migrating monarchs find their way? They fly and glide with little reference to the ground for long periods when the weather, with thermals and high pressure, favors high flight. As for direction, Sandra Perez and colleagues at Monarch Watch have shown what was long suspected: that monarchs possess a refined sun compass. That is, the migrants not only orient toward the sun but also compensate for time by shifting their direction relative to the changing azimuth of the sun as it appears to cross the sky. The scientists proved this by clock-shifting monarchs (fooling them into thinking it was six hours later than it was), then releasing them and comparing their mean directionality with that of normal butterflies. The delayed monarchs headed west-northwest, the controls southwest, showing that their angle to the sun depends on the time the migrants perceive it to be.

Migration scientist Robin Baker, in a paper entitled "When and How to Go or Stay?" wrote that "a butterfly flying across country searching for the next suitable habitat should place a premium on efficient scanning of the ground," while butterflies "predisposed to migrate with the greatest economy to a particular area . . . should [follow] the straightest route from place of emergence to the destination zone." He predicted that such butterflies should have a sun compass. I believe monarchs travel both ways — efficiently searching for nectar and shelter or economically flying in a more-or-less straight line — depending on the need and opportunities of the moment.

I was driving along thinking about this, listening to the radio, when *All Things Considered* reported a new study of bird migration showing that European garden warblers use their genetic memory of the magnetic field as the default system, with stellar navigation only the backup. This is the reverse of how it was thought that they navigated. What about monarchs? Unlike birds and moths that have been shown to use the moon and stars, they cannot migrate by night. Virtually all butterflies, including monarchs, are diurnal. What do they do on cloudy days, and have they any other systems but the sun?

A Mexican monarch conservationist, Fernando Monasterio (a dashing fellow known to his American colleagues as Zorro), proposed that magnetism may be an orienting device for monarchs, pointing out that they home in on the highly magnetized highlands of Michoacán, and they do have magnetite in their bodies. C. B. Williams, the great English pioneer of insect migration studies, designed elaborate experiments to test whether magnetic fields contributed to insect orientation. In *Insect Migration* (1958), he pondered "how to persuade the authorities to provide the necessary machinery for a wild goose chase to a remote part of the world to test the effect of a magnetic field on migrating insects — when it is almost certain that the answer will be negative." As it turns out, the question did not have to be that complicated. Sandra Perez and her colleagues' successful sun compass studies were followed by simple but elegant experiments in magnetism. Monarchs exposed to an intense magnetic field were unable to orient and departed with random bearings — demonstrating that their sensory systems can detect magnetic fields, which vindicates Williams's ideas and shows monarchs to be even more complex than previously thought. My own travels would suggest again and again that they seemed to know where they were going, by the various means at their disposal.

I changed the station and listened to the "Ave Maria" from Verdi's version of Shakespeare's *Othello* as I drove from Odessa toward Othello via Stratford. A little farther on, I found myself seeking monarchs on Urquhart Road. The Urquhart brothers of local note, probably unrelated to Professor Fred Urquhart, whose name is synonymous with monarch studies, had settled between Marlin and Wilson Creek in the 1870s. East of Marlin (also known as Krupp) I passed a pile of baling wire the size of a school bus and a scummy old stock tank with a flora like the Hanging Gardens of Babylon. The ruts died at the railway line in heat and weeds above Crab Creek.

Even if monarchs sometimes travel way up high, keying their flight plan on the sun and maybe the earth's magnetic field, they need to come down to earth now and then for nectar and shelter. Crab Creek, arising west of Spokane in a substantial coulee and wending for two hundred–plus miles across a broad swath of eastern Washington before joining the Columbia below Vantage, varies from a modest waterway to a wandering rut. Crab Creek offers over its long course an intermittent series of stage stops for butterflies.

When I pictured the monarch I last tagged leaving Grand Coulee, I saw her circling on a thermal, rising above the entablature of the coulee basalts, ascending to a height from which the land below could only be a blur darker than the ceiling, and sailing on. She floated on the updrafts, glided on the still air, tacked on the wind, aiming into the perihelion to find south. This she did for hours, until cooling air and the urge to find nectar brought her down on a long, shallow glide. When she neared the surface, she searched for green or for landforms that might mean moisture. I imagined that her faculties are adapted through ancient selection to do this. And what she was likely to find was Crab Creek. That is why I chose it for my route back to the Columbia.

My first crossing of Crab Creek came on a wood-planked iron footbridge in Odessa. The creek bed was broad, having already run for many miles, but bone dry. Maybe boiled dry: the temperature was 97 degrees, and it was hard to imagine that the nights would soon be too cold for monarchs. I briefly parked among the pickups in front of the Harvest Moon tavern and entered its dark cave for a cold Coke. Odd looks shot from the cowboys at the bar. They didn't ask my business and I didn't tell.

I next struck Crab Creek at the Gloyd Seeps Wildlife Area, a place of extensive milkweed and nectar. Walking across the marshes in the predusk, I saw two female coppers roosting under milkweed leaves; one ochre ringlet was bedded down in tall grass, another was an ambush bug's dinner on a rabbitbrush. The air had gone cool at last, the sun set into cloud. Crab Creek crossed the seeps, now fat with water. It boiled through a rumbly culvert from a pond spattered with ducks, coots, and gulls. A bittern flew over, snipe warmed up for their winnowing flights, and a little black merlin scouted past. Marsh sounds: shrill hoots, brittle whispers, wren buzz, zither of mosquito. I glimpse the eye of a rodent as it works the rushes and cattails and leaps between them. Four men in orange vests with shotguns, and one with neither, appeared on the path and fanned into the brush. Later I heard one blast. I made camp in a parking lot littered with shotgun shells. In the south, heat lightning flared. Just before dark, red admirals flew in, strong and southeasterly, to roost in willows.

Next to the monarch, the most famous migrant butterflies in North America are the painted ladies, genus *Vanessa.* Only the basic painted lady, or thistle butterfly, *V. cardui,* moves by the millions, occasionally shutting highways with their very mass. But the other three species undergo spring and fall movements that we understand even less than those of monarchs. One of them, the crimson-banded red admiral (or, more properly, admirable, *V. atalanta*) was the last butterfly I'd seen the evening before. Now another, a fresh west coast lady (*V. annabella*), opened the new day by nectaring on goldenrod in the early sun. For a hopeful second I mistook it for a tiny monarch. The two butterflies share similar colors of oriole orange and black, but there the resemblance ends: the lady is only a couple of inches across and blotched with black rather than sharp-veined.

When I was a boy I saved butterfly ephemera faithfully. One item in my bulging files was a favorite *Reader's Digest* cover showing monarchs and orange sulphurs lazing among the silky seeds of milkweed pods in an autumn field. As I walked out onto Gloyd Seeps again, there was that scene, minus only the monarchs. The idyll cracked as a gargantuan aluminum flier passed over instead. Japanese airliners, B-52s, and other big jets howled past at low elevation every few minutes after that. Nearby Larson Airfield was used for training pilots. One shiny new jet was emblazoned "JAL Super Logistics" on the side. No doubt. But its gleam was dull compared to that of the arcing magpie, its logistics coarse compared to those of a migrating monarch. I was now directly southeast, and not very far, from #81741's release site at Alkali Lake.

I worked acres of largely yellowed and browned-out milkweed surrounded by willows and goldenrod. Just as the *Asclepias* was almost past, so were the mosquitoes, one of the pleasures of autumn pursuits. The air was thick with sweet scents like the soap aisle of a supermarket, but better.

Gloyd Seeps was alive with butterflies. Besides the sulphurs larking about, three kinds of whites crowded the watercress and little sunflowers. A hopeful but tatty male mylitta crescent tangled with a ringlet that flew past his patch of road, while another held a territory twenty yards on. A big female wood nymph darted through the bunchgrass, and when an orange dragonfly distracted me, she vanished into thin air, as wood nymphs will do. Near the marsh an old female field crescent basked. She played possum while I examined her field marks, then flew while I was debating whether to keep her for a voucher specimen — an irreproachable proof for future reference. I hadn't collected anything but the occasional road kill so far on this trip. I generally prefer my butterflies alive but will collect specimens if it seems worthwhile, for example to verify a range extension.

As a young collector I made bargains with my "Butterfly Gods": I'll release *this* if you'll let me catch *that.* Now I proposed letting the crescent go (she was slow, easy to renet) in exchange for a monarch or two. No doubt the gods knew I hadn't the heart to collect a rare female anyway, and ignored me. She ignored me also, basking next to a female copper on Canada thistle. A big rusty assassin bug waited on a goldenrod and probably caught the little checkered butterfly as soon as I turned away. Checking the butterfly atlas later, I found that this species had never been recorded in the Columbia Basin. Without the specimen, I'll just have to go back and find it again.

I did collect something. In the marsh, a glint beneath tall mauve asters gave away a stack of aluminum cans, mostly Lemon Nestea and Cherry Coke, along with an empty package for 9mm Luger pistol cartridges. When I finally gave up on monarchs, I went back for cans, subjecting Marsha to the indignity, not for the first time, of transporting a load of aluminum. Then I pulled up a stand of purple loosestrife, a beautiful plant and a good monarch nectar source, but an alien that overtakes native wetland vegetation wholesale. Picking up trash and yanking invasive weeds were not what I'd signed up for. But once you've been a nature reserve manager, it's hard to stop.

In the parking lot, beer cans full of tenebrionid beetle carcasses shook like Indian rattles. Crushed underfoot, spout down, the cans left a midden on the mashing stone, like the leavings of some occult rite involving darkling beetles anointed with Bud Lite. Altogether I had sixty-nine aluminum cans. At the Moses Lake recycling center I turned over the lot to a Mutt & Jeff pair of trashmen. (There is no deposit in Washington, so I received the actual value of the aluminum.) The short one, wonderfully filthy and Dickensian in his long, greasy duster, doled out my payment penny by penny and looked at me as if I were quite mad for bothering to bring in a cache of cast-off cans. But who knew when I might need that fifty-eight cents? My grubstake for hunting monarchs was minimal.

The only time I'd actually been paid to hunt for caterpillars was here in Moses Lake in the summer of 1984, when I was a caterpillar wrangler for a Hollywood movie named *Runaway,* starring Tom Selleck, Cynthia Rhodes, Kirstie Alley, and Gene Simmons. Directed by Michael Crichton, the film had something to do with errant robots, and an early scene showed farm robots plucking caterpillars from corn until one robot ran amok, requiring policeman Selleck's attentions. Hired to provide one hundred caterpillars, I spent three days on set, sharing an air-conditioned trailer with Selleck's stunt double and stand-in. They compared notes on tequila, gold, and women, and I showed them my larvae, some sixty tomato hornworms and a few anise swallowtails, striped rather like monarchs. When the time came for our scene, Crichton called, "Bring on the bug man." But when he saw the soft green hornworms, he said, "Oh, shit."

I guess he wanted purple and orange. But the director of photography, John Alonso, decided the hornworms were fine, and after a few failed takes, a hornworm cooperated and the scene was in the can. But the director wanted to try one take with a large green beetle sent up from Hollywood. "Get the bug," said Crichton; but I had chilled the beetle in the ever-present beer cooler, and it was almost moribund. While everyone waited, eager to shut down and get out of dusty, torrid Moses Lake, I blew on the beetle to warm it. The third assistant director said, "I can't believe it — mouth-to-mouth resuscitation on a goddamned bug!"

I liberated the rest of the hornworms in Thea's potato patch in nearby Wenatchee. She and her children, Tom and Dory (then ardent fans of *Magnum, P.I.*), had joined me briefly on the set, and Selleck offered to pose for snapshots with them, which greatly raised my stock with my future stepchildren.

The same glacial floods that scooped out the coulees scoured the soil and plucked loose lava from the surface of a vast area in eastern Washington. This left countless depressions, many now water-filled. Called the channeled scabland, this intriguing portion of the Columbia Basin contains hundreds of lakes, ponds, and potholes. As I entered the Potholes, a monarch crossed my bow heading south. The next monarch, a larva opposite a Crab Creek crossing, languished on a dying hank of milkweed sprayed with herbicide — on the Columbia National Wildlife Refuge! I moved it to a healthy plant, and it wandered the leaf up and down. Maybe it had already begun its walkabout, the sometimes extensive ramble that caterpillars undertake before pupating.

As a monarch corridor, Crab Creek was looking good, lined with wild rose, licorice, Russian olive, nectar abounding. I walked the Crab Creek Riparian Trail to a waterfall flowing into a cove of the creek. There I put up two magpies, three pheasants, one grouse, one great blue heron, a dozen splendid wood ducks, and a pied-billed grebe that surfaced and spattered. At such times I feel very rude. The cool falls sang a siren song on the 100-degree air, but I didn't want to disturb the fish, frogs, grebes, and ducks any more than I already had. When I got back to Crab Crossing, the caterpillar was not to be seen. I hoped it was off making the J form they assume just prior to pupation, rather than dying of poison. A Pacific rattlesnake lay dead in the road, with two ornate and stinky carrion beetles in attendance.

Where Crab Creek crosses a corner of habitat-poor Adams County, Thea and I once found a refuge among the ubiquitous fields, a spot rich in life forms such as huge and elegant orb-weaving spiders and a spectacular mating pair of Oregon swallowtails. Now I found the site fenced as part of the refuge, and I could see only cabbage whites and mosquitoes, and carp mouthing oxygen. It was protected, but no longer accessible to the casual rambling naturalist — a common tradeoff as the natural landscape shrinks.

Smells of onions and mint rose from the fields as I regained my guiding watercourse at Corfu, a name with no place, a sometime railroad stop on Lower Crab Creek. As I followed the railroad line down toward the Columbia, miles and miles of rocky washboard disappearing into a dusty sunset failed to erase my high hopes for the stream's home stretch. They would be justified, but not until the sun had gone most of the way around once more. Now, in high winds and darkness at Vantage, where I-90 crosses the Columbia, I rendezvoused with a friend and his dog. I'd met David Branch, a Seattle dentist, in a butterfly class, and we'd since made a habit of taking to the field together each fall. When I asked him to join me on the monarch trek for a couple of days, he dropped his drill, grabbed Karma, and came straight away.

The next morning we wandered south on the west bank of the Columbia, taking to the shore below Wanapum Dam. Karma, a black border collie, nosed up a nest made of a wiry, feathery fiber. I guessed a savannah sparrow had turned the wickedly invasive water milfoil to its advantage. For any insect-eating bird there was no shortage of food; a small caddis fly was having such a hatch that it was hard to walk without inhaling them.

We nosed into a private park at the end of a dammed side canyon that looked nicely brushy. Like Daisy, way upstream, Doris had been a real settlement before the dams. Now it was just a campground, with an odd beaver lodge built into a half-sunken boat. "Monarch!" David called, but it was a viceroy butterfly cruising about in the morning sun. So the famous mimic had entered the picture at last. Having the monarch's stunt double on the scene would complicate matters but also make them more interesting. A little later, across the river, a big female viceroy skimmed the low willows, probably laying eggs, and I was the one fooled, netting her before I realized my mistake. Released from Marsha's silky embrace, she sailed, alighted, and basked on flowering Indian hemp with her wings spread. She was citrus-orange, with white-spotted black wingtips much like those of her model, the monarch, but she also bore a black crescent cutting across the veins of the hindwing. This stripe is the giveaway (along with wing shape, size, and other details); the monarch doesn't have one. Lightly lined with white, the hindwing stripe is a remnant of the aboriginal "admiral band" that adorns the wings of her evolutionary ancestors and current congeners, the white admirals.

Viceroys have come to resemble monarchs just as other species of the genus *Limenitis* (the admirals) have evolved as mimics of other unpalatable models; for example, the red-spotted purple mimics the black-and-blue pipevine swallowtail. These mimic species arose from white-banded black ancestors as natural selection favored offspring that progressively resembled the unpalatable model species. Orange or blue mutations that looked something like monarchs or pipevine swallowtails enjoyed higher survivorship as birds left them alone, and thus lived to pass on their genes, leading to yet oranger and bluer generations. The basic fact of butterfly mimicry was long suspected by H. W. Bates, F. Mueller, Darwin, and others. That it actually works as hypothesized was finally demonstrated by Jane van Zandt Brower and Lincoln Brower's elegant experiments, in which naive bluejays were presented with monarchs and, later, with viceroys. The hungry birds gobbled, then tossed their monarchs, finally appearing abject and educated. The emetic qualities of monarchs sickened the jays before their cardiac glycosides could kill them. Thus conditioned, the experienced birds carefully avoided the similar viceroys.

Conventional wisdom since has been that tasty viceroys mimic untasty monarchs in classical "Batesian" mimicry, meaning that only the model is unpalatable. However, more recent research by Brower and colleagues has shown that viceroys acquire their own bad taste from their willow hosts, thus teaching (and confusing) avian predators just as monarchs do. This makes them "Muellerian" mimics instead, whereby two or more bitter-tasting species play both ends against the middle — the middle being the bird — by resembling one another. As David and I showed, even discerning netmen can be taken in, if "hungry" enough. The viceroy went back to her business in the willows.

"She's so fresh, I'll bet she emerged this morning," I said to David. "Last night's winds must be tough on her kind." He'd caught and released a Juba skipper and was now foraging wild onions for the evening's teriyaki.

"Do you think the wind accounts for there being so few butterflies here, when there's so much nectar?" David gestured to the paintbox borders of asters and coreopsis, gaillardia, loosestrife, and lupine that spread around the sandy peninsulas and shallow embayments we'd been wading.

"Maybe," I said. "Anyway, the breeze is out of the south now — no help for monarchs! Fortunately, they're expert at resisting the prevailing winds."

A pair of geese from a big flock launched from the river, landed on a gravel island, and honked back our way. "Nice to see some geese that still want to migrate," David said. Once the classic symbol of autumn migration and wanderlust, Canada geese have so adapted to urban environments that they have become all but sedentary in many North American cities. Huge numbers of the big gray, white, and black honkers have settled down in Seattle, abandoning their romantic reputation and smearing parks and upscale lakeside lawns with goose poop.

An iron railroad bridge crossed the Columbia River to meet the railbed I'd followed the previous night from Corfu on Lower Crab Creek. By a huge orchard, Hispanic women waited to put their children on a school bus before crossing the old bridge to the store at Beverly, on our side. I wondered if Mexican migrant workers dream of *mariposas monarcas'* ease in simply gliding over the border at whim. But the Mexican monarchs' need is just as imminent, their challenges merely of a different order. Each by its own means, human and insect take on the frontier. Individuals fall away, but en masse, they make it.

David and I turned east, back toward Corfu, to spend the afternoon in the Lower Crab Creek Wildlife Area. Showy milkweed spread out in acres and acres. A mess of viceroys worked the willows, but we weren't to be fooled again. Crab Creek was big and broad here, and full of water, a long way from Odessa. We were directly in line with the Potholes via Crab Creek and due south of Sun Lakes. It seemed logical that monarchs coming from those sites would pass through here on their journey south and that in spring, breeding would occur here among all this *Asclepias speciosa.*

Cabbage whites nectared avidly on purple loosestrife near a purple sign that said this is a "Purple Loosestrife Control Area." We took separate anglers' paths north toward Nunnaly Lake. Mine crossed a cool, artificial glade of old Russian olives. Shed foliage, like straw underfoot, was rich with the smells of autumn harvest. Suddenly a roman candle of red underwings exploded from the dark wood. These big, beautiful moths hide by day against bark, their cloaking forewings perfectly cryptic. If disturbed, as by me, they drop and fly and flash their hindwings — rose-scarlet with black bands — supposedly startling potential predators for a moment while they flee. Okay, I played: "Whoa! Underwings!" And they were gone before I had a chance to net one to show David. Through their own fallibility, lepidopterists go a long way toward demonstrating the reality of insect defense theories.

When we rejoined, David and Karma walked one way along the old railway that made a raised trail through the wildlife area, while I checked the other. At 4:30, overcast, about 70 degrees, the breeze getting fresh, I saw a big monarch in a patch of milkweed on the south side of the road. It was exploring for nectar, slow and unexcited, unlike me. I was entirely focused on the Halloween-colored kite drifting before a backdrop of sand dunes, the Saddle Mountains, and, nearer, the creek. It crossed the road just in front of a speeding camper, paused on a milkweed leaf, then flew to a dried weed stalk and basked. I stalked, fully expecting it to launch a second before I could act. But it stayed. I inched closer, holding Marsha behind me out of sight but above the vegetation, moving like a sloth. Then I changed to a cat, struck fast — and caught her. I whooped, and David came quickly.

The rising west wind made tagging difficult, blowing the wings and the papers about, so we retreated into David's Jeep to do the deed. Monarch #81746 was a perfect female. By 4:50 I had replaced her on a goldenrod. I could *just* read her number at eight feet with the lower part of my blended trifocals and my little Leitz binoculars. She spread her wings slowly, mostly sat folded, the wind tossing her a bit. Karma wanted to see what we were so excited about, but David called her off.

"Maybe she eclosed here today . . . or moved down Crab Creek," I said.

David was grinning. "It's so *much* bigger than a viceroy," he said, holding his windblown brown hair out of his eyes.

". . . and statelier," I said. "The names are right. There's no question when you really see one." As I said that, a *big* viceroy shot by on the breeze (I thought).

Sharing a beer and trail mix, we watched her until six, when we decided to move her to a less exposed spot and check her in the morning. Then David noticed that she was holding her left forewing lower than her right forewing. I had maladroitly stuck her forewing to her hindwing with the tag! It's fairly easy to make this mistake if you're not fully attentive, and the wind had made me clumsy. But it was still embarrassing, and a bad deal for the butterfly if not corrected. I fixed it, and she was fine. We replaced her in a secure spot in the lee of a big Russian olive, six feet up, hanging from leaves, the temperature 72 degrees F.; she seemed comfy, and there we left her at 6:30.

Back at the KOA in Vantage, we barbecued marinated teriyaki steaks with the wild onions, and David opened a bottle of Gossamer Bay Zinfandel, whose label bore a handsome monarch on a grape leaf. A cardboard monarch, very realistic, was looped about the neck of the bottle. It would travel on with me, pendent from the tuning knob of Powdermilk's radio, like a decoy. I hoped it wasn't the last monarch I would see.

After dinner, and pie at the Vantage Cafe, we drove over to check on Ms. 46. She was hanging just as we had left her, but I could see the tag projecting over the left hindwing, which wasn't right. Checking, I saw that when I had adjusted it earlier, part of the tag had not restuck well, having gotten scaly. It could get in her way. So, with my Swiss army knife scissors, I cut off the part that says "mail to Entomology." The rest, with "Nat. Hist. Mus. LA, CA 90007" and her serial number, remained intact. She sat tight as the wind rose.

That famous Columbia Gorge wind blew my sleeping bag and pillow away before I could get in. Buck naked, I had to recover them in the dark, fifty yards away and full of dust. I lay on the KOA lawn thinking of Ann Zwinger, in *Wind in the Rock,* as she tried to sleep through "a blue screamer" of a desert windstorm, "secure in the knowledge that tomorrow is likely to be a bummer." I didn't feel that way. But I did wonder how one is supposed to follow a bit of tissue on a hurricane, whose big idea this was, and how that gorgeous, much-handled monarch was doing out there in the gale.

We got back to the site at 7:30 A.M., and she was still cozily roosted. At 8:00 it was 61.7 F (we'd both guessed 62). The day was mostly cloudy, with shots of sun and sprinkles, and only a light breeze still out of the west. I swear there was a rainbow over the monarch tree after a shower. The buzz of the power lines resonated with that of the few mosquitoes. To the south rose the huge brown-and-tan-striped mass of the Saddle Mountains — a frisky morning's hurdle-jump for a departing monarch, unless it dropped down to the Columbia gap. At 8:30, it was half a degree cooler, and she still hadn't moved.

We shared coffee, cantaloupe from Daisy, and some of the last of Thea's granola cut with Kix. A rockslide rumbled in the basalt behind us, the wind shifted easterly, 5–10 mph, the air warmed to 62; at nine, 63 and climbing, sun struck the tag. She moved a leg with the next sunbeam. The cloud mass was shifting easterly, promising full sun soon. A big green darner alighted nearby in the Russian olive; it better not have any ideas, I thought.

Number 46 clung, wings down, head up, to two olive leaves. One of the reasons that California winter monarchs have been able to make such a successful adaptive shift to eucalyptus groves, in the absence of native conifers, is that the leaf width nicely fits their grasp. The long, narrow blue-green olive leaves were similar to euc leaves, if smaller and softer. Russian olives having replaced the native willows over vast areas of the riparian West, I concluded that a similar adaptive shift had taken place here for overnight roosts. Magpies chortled shrilly, and pheasants cackled, as David said, like an old Model T.

Full sun at last, and the first wing spread, at 9:17. A drop of meconium, waste from the pupal process, fell from the tip of her abdomen, suggesting that she had indeed eclosed here yesterday. It was dark, glossy olive, not like the coral-red meconium of painted ladies and many other species. Nabokov and others have suggested that it was the bulk drippings of meconium during mass emergences leading up to great butterfly migrations that gave rise to biblical and other apocalyptic accounts of "rains of blood" in ancient times.

At 9:19, 67.3 F., breeze from the north at 10 mph, she preened her antennae, spread her wings, and left them open to bask. And then, at 9:20, 68.5 ("that degree they recognize," as a friend has put it in verse, when "the sun's heat ignites them to flame") she flew. I was ready to sprint after her to track her direction, but she merely moved over a few feet to a branch in full sun and basked with her wings closed. It was her first free flight since I'd netted her seventeen hours earlier. Five minutes later she floated again, and, as we watched with open mouths, prepared to start jogging behind wherever she bid, glided to the next tree, about twenty yards to the east, and settled ten feet up. The wind freshened from the NW, maybe deterring her flight.

David waited until 10:15 to leave; he would have to drive like hell over Snoqualmie Pass to meet his appointments. At 10:30 our monarch was still enthroned. Clouds had come back. "You might be there all day," I whispered, and she closed her wings. Her short flight had been controlled, flashing, unimpaired. I lay in dried grass and milkweed beneath her, in an autumn idyll.

I wondered if this could be a long low-pressure sitting. Urquhart wrote of someone in Utah being unable to *make* monarchs leave Russian olive trees during a spell of low barometric pressure. I declined to try. A blue grouse may have been booming nearby, but it was hard to tell, because ATVs had begun roaring like bloated hornets in the sand dunes across the creek. The dunes are known to support endemic insects, but the state Department of Natural Resources had designated the area for dirtbike and dune-buggy use regardless.

Dappled sun, and she spread, moved a little. I asked myself, What concatenation of urges does she feel? Is there anything like desire beneath that chitinous cuticle, or is it nothing more than the expansion and contraction of various tissues as prompted by the temperature and day length, as dictated by the program in her genes? "Desire" is a strong word, and few things are more subject to human projection than creatures with which we identify. Perhaps the tangency of urgencies within a monarch that is READY TO GO is nothing more than the sum of its electromagnetic impulses. There may be nothing at all mystical about it — but mysterious, yes. For this is a complicated organism, a supple beast.

At 10:36 she moved around her limb to a more protected spot where she would not be buffeted by other branchlets in the breeze. I dropped my eyes to rest my neck. Ants and wasps covered the cantaloupe rind, and a giant, red-berried, milkweed-seeded, yellow asparagus frond nodded in the stiff breeze. I looked back up and . . . she was gone!

For nearly half an hour I examined every twig, every leaf, in the vicinity of the last sighting. At last I decided that no. 46 had either flown while I was looking down to make a note or was hunkered so deeply and well that I would never find her. Every folded Russian olive leaf began to look like a butterfly. I was reluctant to leave the first monarch whose overnight bivouac I had (more or less) shared. But feeling my own urgencies, and the days growing shorter, I finally departed downstream.

Noon found me at the mouth of Crab Creek, where this highly attenuated watercourse of very low gradient finds the Columbia River at last. Absent first Thea, then David, now Ms. Monarca (who gave me the slip), I felt a little bereft. The flip side was being really free, in the field, for a long time to come. As Bill Kittredge put it in *Hole in the Sky,* I was "out, away to the world with hope."

FIVE

Columbia

South of crab creek and Wanapum Dam, the Columbia River plunges through the deep and narrow canyon known as Sentinel Gap, then flattens out. As the major route in and out of the interior Northwest for everything from salmon, geese, and loons to Lewis and Clark and barges, the Great River of the West conducts the butterflies of passage to and from the basinland that bears its name. In post-flood, predam centuries, human settlements sprinkled these productive shores. They still do, but they are very different from the Indian camps and from one another. A nice big shore habitat looked inviting from a distance, but pulling in, I found it a linear, litter-strewn homeless camp built mostly of torn and flapping plastic tarps. Men sat on car seats in the dust, smoking. I caught a young woman's eye, and it seemed to carry a rainbow of regrets.

Then came Desert Aire, a planned community, mostly trailers and prefabs tucked tidily above a sandbar willow shore. People worked on their lawns in the hot wind. A woman looked visibly relieved to see me go. On a third flat, trucks and trailers (no cars) congregated on uneven cobbles and deep sand. Daughters and wives comprised a temporary women's village where meals were made, games played, and interloping males and fish duly admired. The fish were accepted, the men sent out to fish some more or to trailer the boat for the haul back to the Tri-Cities, Yakima, or Goldendale. In these latter-day encampments of aluminum, plastic, and lawns, I saw wrenching reflections of other villages belonging to people and fish now absent except in shrunken, diminished versions of, as the treaties put it, their "usual and accustomed places."

The broad span of Vernita Bridge took me over to Riverside, where I turned back upstream. A scratch of a lava track passed below dramatic basalt formations pocked with thousands of cliff swallows. The lava flow crumbled up into caves, hoodoos, pillars, and columns in a palette of browns and a confusion of shadows, spattered with bird lime. Stands of an autumn-blooming buckwheat looked like snowdrifts, as its name, *Eriogonum niveum,* implies it should. Down by the water, mulberries and Chinese elms overhung the beach. I noticed the green hearts of *Viola odorata* leaves among the milkweed and other native plants. These sweet-smelling garden violets and the exotic trees told of a homestead that once flowered here. Originally the homesteaders would have confronted annual floods in spring. Once the big dams were built for hydropower, like the Priest Rapids Dam that lay just upstream, the regular, renewing floods were controlled. Now the violets, milkweed, and all the creatures that feed on them fluctuate with the engineered outflow and the thin silt on the river cobbles.

In a milkweed leaf rolled with silk, I watched a gray-and-black hunting spider with a yellow blaze on her abdomen, squat and alert. I'd noticed many species of spiders living in and around milkweed, a mark of abundant traffic in insect life. A tan, russet, and beige silver-marked moth with a fancy thoracic furpiece hid under another leaf, head down near the base, and a black, yellow-saddled, tussocky tiger moth larva browsed nearby. Monarchs share the floppy leaves with these and other insects adapted to feed on milkweed in spite of its toxins. Even humans consume milkweed, boiling the young pods to remove their bitter taste. I doubted that this was done now in any of the settlements I'd seen across the river, but both Indians and white pioneers did it over much of the continent.

Three big monarch larvae seemed to leap out at me, then a fourth. I wandered along the bank, and an adult monarch sailed into view from upstream. It perched in an elm, sunned, flew down to a grove of little mulberries on the beach, turned and floated back upslope toward me. I caught her on the wing, another very fresh, perfect female, tagged her, and put her back on the elm where she had alighted briefly. She crawled up a bare limb, basked, pumped, then sailed up toward the road and the slope. *Just then,* vast swarming clouds of caddis flies arose, thousands in each congealing nimbus. Both caddises and the sun were in my eyes as the monarch, magicianlike, disappeared behind a smokescreen of insects and sagebrush.

Sweeping a netful of the little caddis flies and nabbing one for a look, I saw that the translucent gray wings were about a centimeter long, the fine antennae a little less. The aquatic larvae of Trichoptera construct protective cases of pebbles, sticks, or other detritus in which to live on streambeds and pond bottoms. I wondered what the larval cases of this species were made of, and whether the entire bed of the Columbia was covered with them before the hatch. In their mating swarms they shift together like sandpipers in flight, responding to one another's minute movements. Backlit in the sun they looked like swirling snow, as if all of the buckwheat petals had taken off together and come alive. They boiled over the black slopes, recently burned even blacker than the basalt, in uncounted millions or maybe billions. It might be that I had never seen an animal this abundant, even the impossible mobs of monarchs in Mexico.

A huge darner dragonfly rested in a nearby sage, her abdominal spots the same pale turquoise as the pungent sage buds. A great blue heron stood on a point silhouetted against the river. But the monarch had vanished. She was clearly working her way southward along the shoreline when I first saw her, and I guess she continued. I turned around to follow that way. Starlings burst out of the basalt caves as I started the engine. A chukar squawked as the still-warm sun melted behind cold lava, reminding me of a local confection called Chukar Cherries. The track ran south through cherry orchards. You could tell without seeing the trees that it was cherry country from the abundant coyote scat full of cherry pits. I was hungry, hot, and thirsty, and wished I had a bag of chocolate-covered Chukar Cherries and a big icy glass of cherry cider.

Yet I also felt renewed, recovered from the hot wind, the sad riverside camps, the partings. A monarch a day keeps the blues away. The Columbia rolled on, a broad, sluggish thing here, waddling between rimrock and gravel flats, orchard and dam and powerline. Below

Vernita it stirred with the suggestion of a real current as it lolled toward Hanford Reach: the only undammed stretch, the real river, and my next stop. Virga, those gray wisps of evaporating rain cloud, stroked the southeastern sky like the eyelashes of the gods.

No highway runs along Hanford Reach, one of its glories. A bleak drive over the plateau and back down a grade brought me to Ringold in the dark, where Powdermilk just fit on a knoll beneath a sheltering alder. Swarms of midges and caddis flies completely covered the windows. At four in the morning, the sounds of guys in trucks and boats pulling in woke me. I turned over in my comfortably reclined car seat, shifting weight to the other cheek and tugging my sleeping bag around my ears.

Butterflies being quite civilized, there is no need to rise at break of dawn to see them, as one must for birds. Cloud and cold wind met my next awakening, but on the third I saw cabbage butterflies and their shadows, and I stirred. I was perched beside a rushing canal, a wasteway for Potholes water from the plateau into the Columbia, mainlining agricultural chemical runoff into the Hanford Nuclear Reservation's radionucleide brew. Hanford has dominated the Big Bend of the Columbia River for more than half a century and has left parts of it as polluted with radioactive wastes as anywhere east or west of Chernobyl. Reactors and their fell smokes and red steam hulked on the far shore, but I was not interested in thinking about it much that morning.

What I was interested in were Hanford's inadvertent treasures. Rattlesnake Mountain rose across the river to the west, showing its long, shady blue side. Lupines had painted much of the mountain indigo when I first saw it the previous spring, the summit abuzz with hilltopping swallowtails of four species, about the time north-flying monarchs were returning. I was one of a group of ecologists visiting the Arid Lands Ecology unit (ALE), a research reserve that has become a great unlooked-for beneficial effect of the government's occupation of the region for the atom-bomb-building Manhattan Project. ALE, the largest piece of eastern Washington shortgrass steppe left unplowed and ungrazed for the past fifty years plus, is a site of enormous ecological significance.

The other treasure is Hanford Reach. Had it not been for the reactors and their appropriated hinterlands, the Ben Franklin Dam might have gone in long ago. Now that the age of giant porky water projects has largely passed, and the removal of certain dams has been officially mooted in the debate over salmon survival, it seems unlikely that a dam joining Benton and Franklin counties will ever be built. Now, except for the tidal stretch where I live, this is the one place where the Columbia runs wild in Washington. I'd guessed that it would offer prime monarch hunting grounds.

The dirt road north from the Ringold hatchery follows the river below Wahluke Slope through public wildlife lands. Gravel spurs lead to the sites of old homesteads above the shore. I took the second one down to a shady grove of honey locusts, and, climbing out of the car, I beheld a great big beautiful monarch flying about a locust, then another down among the knapweed. For the first time I was tracking two monarchs at once! The first sailed to another locust, and I missed a high shot. The other rose out of the knapweed, flew south, then back. I lost it but caught a third. Tagged, she flew off fast at one hundred feet plus, southeasterly, toward Wahluke Slope, as a swallow made two abortive passes at her.

Surely some monarchs must be sacrificed to the swallows' education. After demonstrating that blue jays really do avoid monarchs once they've been subjected to their bad taste, Lincoln Brower went on to show how certain grosbeaks and orioles have taken to preying on the massed monarchs in Mexico. The grosbeaks simply scarf them down, buffering the cardiac glycosides with other foods. But the orioles actually pluck out the untainted, nutritious fat deposits, dissecting the prey with their sharp bills. I have watched them in the forests above El Rosario in Michoacán, four species of orioles in large flocks the colors of the monarchs themselves, stripping butterflies out of the firs like kids in a berry patch. When that much food is concentrated in one place, some organism will usually find a way to make use of it. Orioles figuring out how to feed on monarchs are not unlike the monarchs themselves evolving a way to use the vast milkweed resource of North America. In both cases, animals have come up with means of getting around protective toxins in order to exploit the goods within. With monarchs more dispersed in the North (especially the Northwest), birds have less incentive to find a way to eat them, so they generally do as they are taught by parents or by direct experience. But it does take a few monarchs to do the teaching.

The sun came out and meadowlarks sang as I set up my stove for a bite and coffee. Then another *Danaus* appeared on the upstream side of the grove, and I lost it down the bank into the blue willows. These butterflies had evidently spent the night in the beige-leafed old farm locusts, which stretched to fifty or sixty feet in height. Still another appeared by the windbreak, and I caught her on the wing and tagged her. Released onto a yellow coreopsis, she flew below the bank, south into the willows before I lost her downstream on the slough behind Savage Island.

The riverside vegetation was composed chiefly of sandbar willows *(Salix exigua),* which I call blue willows for their long skinny leaves, glaucous and redolent of sandy, damp places. A small fresh male viceroy leapt up from the same thicket. Blue willows being one of the favored food plants of *Limenitis archippus,* that's where he was likely to find a female. On the ground, an incredibly brilliant blue-green hunting wasp prowled for a cricket, which she would sting and stuff into a burrow for her single larva. Her shimmer recalled that of the metallic blue Vespa I drove among the blue willows of my youth, hunting viceroys and monarchs and great gray coppers.

What is it with metallicism in Nature — like the gold spots on monarch pupae or the wholly gold chrysalides of the black-and-blue milkweed butterflies, called crows, that we had seen in the butterfly house in Kelowna? *Chrysalis* means a gilded box, more or less. But is all such brilliance solely for the benefit of educable birds? Many beetles are metallic shades of blue, green, gold, and silver, including flower, flat-headed, blister, and tortoise beetles. This hunting wasp has numerous iridescent and metallic kin. Certain species even effect the candy-apple red I once tried (very poorly) to paint my Vespa. It can't be for purposes of sexual selection, since both males and females shine.

At least for fritillary butterflies, with their quicksilver spots, I have heard a plausible theory — that they resemble dewdrops sparkling in the grass, thus conferring cryptic protection upon butterflies warming up to fly in morning meadows. Perhaps this could account too for otherwise camouflaged brown and green pupae bearing dramatic medallions of seemingly molten gold and silver. But I have heard no overall theory of intense metallicism in insects. It may be that warning, of either bad taste or sting, drives such flagrant coloration; or it may be a byproduct of other processes, with no special function. Sometimes we are too quick to seek a purpose for every feature in nature, when many traits, if they confer no liability for survival, may be fixed simply as side effects of genes with entirely unrelated tasks to perform.

By noon it was clear, in the seventies, the wind blowing only off and on. The river here was as fine a corridor as I had hoped it might be, full of nectar — mauve aster, yellow coreopsis, purple loosestrife, goldenrod — and milkweed, too. A dragonfly basked on a stone: a solid pale blue libellulid with black vein marks at the tips of all four wings. Its eyes were a deep-spotted turquoise. Just as birders are discovering butterflies, several butterfly watchers I know have taken a keen fancy to odonates — dragonflies and damselflies. I am falling for their charms myself after years of dumb admiration. There were hundreds of meadowhawks, darners, skimmers, and bluets here. The colors of odonates are maddeningly fugitive; to appreciate their hues, you must see them alive; unlike butterflies, whose colors remain intact on the pin.

I was examining a yellow composite I didn't know in the damp swale above the shore when an untagged monarch popped up and paused to nectar on goldenrod. When I stroked long and bricked it, it took off *up* and east, toward Wahluke Slope. Then, admiring golden sulphurs on lavender asters, I saw a big male monarch nectaring on the same kind of aster, closer to the ground than usual. This time I took better care and netted him cleanly. After affixing a tag to his left forewing, I compared him to the little viceroy I had safely tucked, alive, into a paper envelope earlier for the purpose of photographing it next to a monarch. This monarch was *immense,* fully four inches in wingspan, the viceroy less than half his size. The difference is not always so great. An orange butterfly crossed my bow sailing NW, low. I called it a big female viceroy, then almost convinced myself it was a modest monarch; but she flew to the swamp, perched among the peachleaf willows, and I saw I was right in the first place — a big, intensely colored *L. archippus.* When I released my captives, the small one darted right into the willows after her, while the big one flew strong and high toward the southeast.

This was the first male I had tagged since the Similkameen. Unlike many butterflies, male monarchs tend to be larger and paler than females. But the real giveaway is the set of androconial (or sex) patches, black velvety pads of specialized scales also called alar (wing) pockets, located along a vein on the male's hind wings. Specialized cells within the pockets produce male pheromones. In most danaiines (milkweed butterflies), the males possess brushes called "hair pencils" that they can extrude from the end of their abdomen. They wipe these brushes in the alar pockets, then waft the air with molecules of pheromones that help induce the females to mate. Both the hair pencils and their use are much abbreviated in monarchs, but they retain the black patches. Female danaiines produce their own pheromones in their terminal abdominal segments. We know them by the absence of alar pockets and by the broader black scaling along their veins. Since I was finding females almost exclusively, I began to wonder whether male monarchs migrate earlier, on average, than females. Certainly in many species, notably among the fritillaries, males emerge days or even weeks prior to females. And since the southward migrants are undeveloped sexually, there is no reason for fall males and females to stick together.

Eating a honey and raisin sandwich for lunch, I thought of bees. There was a great deal of nectar here, but not many honeybees, whether because of the mites that have devastated them lately or simply a lack of local hives. There was an abundance of native pollinators, such as solitary and carpenter bees and an array of wasps and flies. As Stephen Buchmann and Gary Nabhan have shown in *The Forgotten Pollinators,* when natives decline, plants come to rely on the pollination services of European honeybees, and when *they* drop out, the plants are left high, dry, and unpollinated. Agricultural sprays, crop monocultures, paving of habitats, and other causes of pollinator loss do not intrude on Hanford Reach, so its native pollinators are largely intact, including the monarchs. You wouldn't know it here, but the availability of nectar is clearly a limiting factor for migrating monarchs in many areas. They serve as effective pollinators all along their routes.

At parking area no. 7, another old farm site, I walked along a natural shoreline of cobbles and mud. In a lush little aster-and-daisy swale, I thought I saw a checkered white, but in Marsha's gentle folds it turned out to be a great white skipper, the first I'd seen on the trip. A male, its silky wings shone with the same mother-of-pearl as the opened valves of the green-backed river clams that lined the water's edge. Actual whites, a pair of copulating cabbages, flew by for comparison. The smaller male carried the closed female, and they were harried by other, mateless, males.

The foreshore came down in an asparagus-mulberry savannah. It looked promising, and as I strode into it, a moose of a monarch glided past. I followed at a trot, but it pumped south in a hurry, so I watched it with my binoculars for several hundred yards farther along the bench before it blended into the background. Savannah sparrows owned the benchland, where deer bedded down and red flickers clicked. I avoided the rising wind by walking below the rim of the old river terrace. Now and then, when it seemed there were no more monarchs to be seen, one popped up out of the asters over the bank or materialized beside me on the wind. Depending on my alertness and the gusts, not to mention the animals' keen reactive grace, some I netted and some I didn't.

The final side track lay opposite the one-time village of Hanford, which gave its name, land, and identity to the nearby nuclear establishment at the time of the Manhattan Project. From there I could see the famous White Bluffs stretching away to the north like a regiment of Sphinxes' paws. As the sun settled toward Rattlesnake Mountain, they deepened from chalky white to old gold. On the way back south,

I pulled again into parking area no. 5 because its two big elms looked like a monarch spot to me. I walked upriver to a high bank and sat on a mat of cheatgrass straw, scanning the sunflowers below. A green, flowery sward opened out toward the river through the backwater outlet that, during high water, makes Savage Island an island. Life had settled down out there, getting ready for evening. But when I got back to the car, I looked up and saw two monarchs still in action.

Circling the tall elms, they acted like a premature male pursuing an uninterested female for two or three minutes. Then they both began poking about the foliage for suitable roost sites. The female perched low, and I was able to net her. The swoosh put up the male, which took a higher perch where I could not make him out. Instead of rejoining him after tagging, the female flew high to the east-southeast. I watched her until she had nearly reached the rim of Wahluke Slope, then lost sight. I was sorry she had forsaken the communal roost, for that is what it surely was, even if they were the only two there. I hoped she found another suitable bivouac uphill, or downriver.

Weary, hungry, and well satisfied with Hanford Reach, I headed in search of a roost of my own. The monarchs I'd found might seem like a miserly few in Minnesota, Maine, or Mississippi, but they were a respectable total in the intermountain West. I'd found the first group roosts I'd ever seen in the Northwest. The reach was a good place, for all it had been through, a rich and volatile place where signs beside great klaxons warn "Listen for siren sounding for about three minutes followed by an audible message: LEAVE THE RIVER IMMEDIATELY." The only siren I heard was the high-pitched, tiny one of the rising caddis swarm. I quit the river reluctantly, deliberately.

Two cities often arise where rivers meet, so at the merger of three rivers, three cities make sense. Kennewick, Richland, and Pasco (the Tri-Cities, Washington's fifth largest conurbation) straddle the deltas of the Yakima and the Snake where they debouch into the Columbia. Monarchs flowing out of the milkweed-rich breeding grounds in the Yakima Valley would come through here, as would those of the lower Snake and the ones I'd been following from the north.

At the Yakima Delta Wildlife Park, a maze of trails wove through

Russian olive thickets to the Yakima River proper, where the milkweed was well chewed. I have found monarch larvae up the Yakima Valley as far as Naches, at the eastern base of the Cascades passes that climb past Mount Rainier. Autumn emergents there travel a valley of vineyards and hop fields that smells now like Concord grape juice, now like a brewery. The river route delivers butterfly arrivals into a gauntlet of sprawl and a hydra of freeways that must be negotiated to reach the Columbia. It is good they can fly high when they feel like it. The delta park might offer an oasis before the city passage.

Each path led to a fishing hole marked with blue chicory and fishing litter. The "park" was completely undeveloped, but it was full of wonderfully woodsy glades and glens for a kid to swing a net in without the restrictions of a formal preserve. It seemed like a fine place for fishing and fooling around. Whites and skippers danced across open fields surrounding a lone cottonwood, whose huge bole was spray-painted with crosses, peace signs, "Anarchy," and so on, each vandal in his own twisted way recognizing a holy site and marking it as such. The place went on and on, thick and thrashed, with little nectar. All the common, weedy butterflies of late summer showed up, but not the King.

At noon I crossed a causeway, lined with black people fishing, to Bateman Island, a thumb stuck right into the confluence of rivers. Ten thousand cabbages, a big mourning cloak, a viceroy; goldenrod, milkweed, and limitless roost trees. Could monarchs be far away? Great old Russian olives, red-trunked like cypress, tangled like vine maples, made loops and roofs of interlaced branches, caves and grottoes that would have thrilled me as a kid. They gave me new respect for a plant I knew in childhood as a backyard hedge. My brothers and I shagged the hard berries at each other but had no mature spinneys like this for forts. Russian olive, the Eurasian *Eleagnus angustifolia,* was introduced to the West for windbreaks. Escaped and naturalized, the silvery-leaved trees have had several decades to get this big. The banyan-like interiors of the groves struck me as perfect for hobo camps in this railroad town. Indeed, signs of a big empty homeless camp sprawled nearby.

A man was poking about for Indian artifacts, despite a sign warning of up to five years in prison and a $100,000 fine for taking or disturbing cultural materials: "surveillance is Conducted." Scott showed me two rough-worked scrapers, then put them back. Thin, in a blue shirt, blue jeans, and a blue billed cap that said "Alaska," he had parents nearby but was living out of his truck just now. "I'm what I call a naturalist-preservationist," he said, and told me about a "whole culture" he'd discovered near Lake Owyhee in southeastern Oregon, complete with animistic carved agates, temple floors, and so on. "I have an ethics problem with what to do with it," he said. "Maybe I should just leave it be." He'd taken an agate effigy home for safekeeping: "It was just layin' there on the ground. Somebody would've just took it and sold it."

Noticing my thong necklace with a catlinite turtle and beads, Scott said, "I used to have a necklace like that, but I can't wear it anymore. The spirit world is so close to me now all the time." He was distressed to think of the original village of the island now displaced by the immense middens of plastic, fast-food packaging, and decayed bedding of the camp. Across the water, fancy homes faced the homeless slum: cold comfort either way, I thought, when the folks are home and regarding one another's situation. Five white pelicans sat on delta mud with gulls and a logful of cormorants. I walked off the tip of the island and stood in the cool mingled waters of the Yakima and Columbia rivers, ignoring the radioisotopes and insecticides seeping into my knapweed-scratched feet as I toed a four-inch river clam shell.

A particularly pretty viceroy cruised the sandbar willow edge and settled with its cinnamon wings spread flat on reed canary grass, then flew out to investigate a bright hoop of mating darners. Viceroys and other admirals assume posts, such as sunny branch tips, from which they conduct sorties at anything that might be a potential mate or a competing male. Scott saw the viceroy too. "You know, they're pretty similar to monarchs," he said, "at least the birds're s'posed to think so." My eyebrows must have gone up a notch.

"Do you ever see any monarchs around here?" I asked.

"Oh, yeah — all over. I saw some monarch caterpillars a while back over at Two Rivers Park, on that patch of milkweed near the boat ramp. And I saw a woolly worm yesterday," he said. "Time for trout fishing!"

"Do you use woolly bears for bait?"

"Nah, I move them off the road. But when I see 'em out, I use a woolly worm fly. By the way, I saw something in the paper about some lady doing a study on a little yellow butterfly that's coming north, like killer bees. Doing her Ph.D., I think it said."

"*Really*?" I asked, both surprised and skeptical.

"That's right," he said, "in the *Tri-Cities Herald,* two or three weeks ago." But at the thought of someone doing a Ph.D. thesis, he grew pensive. "If only I could get the pros interested, I could bring that Owyhee culture to light. But they'd probably trash it. I'm going to write a book on it, if I can just get started." He invited me to visit his folks' house to see the agate head, which he was keeping there until he could decide on a proper repository. I was tempted. But the woolly worms were out. It was time to be looking for monarchs.

On the way back to my car, I watched a red admiral alight on a willow branch, suck sap, bat yellowjackets, and fly off with quick flicks of wing. As various brush-footed butterflies often do, it had flown directly into the right part of the tree, searched for a few seconds to find the exact spot, and proceeded to drink at the sap flow. Sap is an important source of sugars for hibernating butterflies awakening in flowerless February or March and for migrants like the red admiral, but I have never seen a monarch visiting sap.

That sight made me thirsty and I stopped for a V-8 at a mom&pop, opposite an interpretive sign for the Lewis and Clark Expedition. Bateman Island was as far as they got up the Columbia River, on October 17, 1805, on their way home. They found Wanapum Indians drying salmon: "multitudes of this fish," recorded Clark, "almost inconceivable." The water was so clear they could see them at fifteen to twenty feet of depth. Clark shot a sage grouse, forty-two inches in wingspread. Absent from the journals is any mention of Russian olive, mulberry, locust, knapweed, purple loosestrife, freeways, nuclear reactors, dams, etc., etc. Almost inconceivable is right. There might have been monarchs, but the journals are silent on the subject.

My next stop was in familiar territory. In the late sixties, my former wife JoAnne and I took a winter field trip to Columbia Park in Richland with a collector friend, Dan Carney. It was a cold day, with the sycamore balls blowing on their stems, and, in the willows, the hibernacula of viceroys quivering in the river's bitter breath. Dan showed us how to find the little sleeping bags, bits of leaves rolled about the center, the midvein poking out and giving them away to the sharp eye. The minute larvae lash their leaves to the branch with silk before retreating into their leaf rolls to hibernate. In spring, with the new growth of willow, they creep out to continue feeding.

Now Columbia Park is all things to all people, with all manner of recreation and far fewer willows. Fortunately, a natural area remains, managed by the Lower Columbia Basin Audubon Society. I found the nature trail and took the path designated as the Monarch Loop. Here, I thought, if anywhere. Sandhill and woodland skippers visited the goldenrod among robust, six-foot-tall milkweed. The underside of one big floppy leaf was covered with brilliant yellow-orange aphids, like bee's pollen baskets. Deep blue berries of Virginia creeper hung on red stems among ghostly Russian olives. Eastern fox squirrels and California quail skittered and whistled in the underbrush.

A blue-green graffito on an observation blind announced simply "Butterfly." From its tower I could see sachem skippers nectaring on thistle, making this a three-hesperiine walk — three species of grass-feeding tawny skippers. The thumbnail-sized golden triangles darted about the glade — *skipping.* In male tawny skippers, the androconial patch is called a brand or stigma. I can see how the impressive stigmata of this skipper inspired the name *sachem,* which means chief. Its rectangular black sex brands, bigger than a male monarch's alar patches, are truly chieflike regalia.

As I was thinking this, as if to illustrate the point, a fresh male monarch descended from the sky onto the goldenrod patch, surrounded by tall wild rose and nettle. I climbed down from the tower, inched past nettles into the thicket, and caught him. As I was tagging him, he reached out with his front tarsi, grabbed my mustache, and hung on hard. So as not to hurt the hooked legs, I allowed him to launch from his perch at will. For a few moments my face was the habitat of a monarch at large. It tickled like crazy, and I tried not to twitch. Finally, with a lift-off flutter so light I wasn't quite sure I felt it, he sailed off easterly into still air, rose above the trail and the trees, and headed downriver.

I prowled south on the Columbia River dike, solid with blooming rabbitbrush, and arrived at Two Rivers Park, opposite the mouth of the Snake River. This was where a lepidopterist friend, Patti Ensor, had once known a roost tree, and where Scott said he'd seen larvae earlier in the season. I quickly found milkweed beside the boat ramp, where a man was trying to start his outboard motor. The plants were caterpillar-free. Nor were there any milkweed longhorn beetles, but some beautiful yellow-on-black, chevron-striped longhorns occupied the goldenrod. I surveyed suitable locusts for roosting monarchs, but Patti's tree did not reveal itself. The only orange butterfly was a slick little west coast lady, bagging some beams as the sun came out for the end of the afternoon before sinking ruddy in a cloud bank. Backlit caddis flies swarmed and swirled over blue willow going gold. There were five days of summer left.

An hour later, the same guy was still trying to start his boat motor. He was skinny and young, with an obese, red-haired wife and two cute little daughters, one redhead, one brunette. The mom was eating chocolate chip cookies while the girls played beside the "Caution: Barge Waves — Do Not Leave Children Unattended" signs. They kept saying "Dodo! Dodo!" The man's wife walked off the dock, tripped on the ramp, and almost went into the drink, dropping a cookie in the action. The little redhead, already pudgy, went for it. As his females drifted off to their truck, the sad dad struggled on into the dusk with the recalcitrant engine.

Curious about the article that Scott had mentioned about "a little yellow butterfly coming north, like killer bees," I sought out the Kennewick library when I left the park. The only little yellow migrant butterfly recorded in Washington was the dainty sulphur, also known as the dwarf yellow or *Nathalis iole irene*. I discovered it in the southeast corner of the state in 1975, and it hasn't been found in the region since; it couldn't be that. No, the article turned out to concern the very butterfly I'd been watching earlier when the monarch had come along, the sachem skipper *(Atalopedes campestris)*. It is neither very yellow nor anything like the Africanized bees, but it is at least moving northward.

The sachem is another species that I had the privilege and thrill of finding for the first time in Washington State. It moved up the Willamette Valley of Oregon in the seventies, and my friend Maurita Smyth had spotted one in her Portland butterfly garden in 1986. So each fall I scanned open spaces and marigold gardens along the Columbia in search of the first Washington pioneers. When I finally found them, it was in a mall.

Thea had business in an office building on the edge of Vancouver Mall, across the Columbia from Portland, on August 28, 1990. As I waited for her, I was watching the many woodland skippers in the garden around the building's flagpole when I thought I saw a female sachem — bigger, longer-winged, more richly colored and strongly marked than the others. But I had no net with which to confirm the state record, Marsha having stayed home. I borrowed a length of stiff insulated wire from a handy electrician and quickly fashioned a mini-Marsha from a little cotton pecan bag. On my hands and knees in the garden, all eyes at the windows above upon me, woodland skippers chasing away the sachems every time they landed, I finally succeeded in netting both a female and a male to establish *A. campestris* as an official member of the Washington fauna.

Since then, sachems have spread along the Columbia to the Tri-Cities, where they are now common. A University of Washington graduate student, Lisa Reed, had been investigating their range expansion in hopes of catching clues about how organisms adjust rapidly to colder climates. Animals extend their ranges northward as the climate ameliorates or as they adapt to colder winters. It remains unclear just how far north, and in what stage, sachems are able to overwinter, but they are expanding from what was once a Sunbelt residence with only summer excursions northward. So Scott was partly right. He was more accurate than the reporter, who called the sachem skipper "rather drab . . . mundane in appearance: brown, with a dab of white." In fact, it is a delicate and elegant blend of golds, olives, wheats, and mauves, along with those startling black brands, like epaulettes of rich velvet.

In any case, it was nice to know that someone else in the area cared about such things. In a town whose high school athletic teams are called the Bombers, where the microbrewery makes Atomic Ales, and where nuclear waste disposal is the hot topic among the morning latté crowd in Jennifer's Bakery, you don't have to dig deep to find most folks' agenda, and it isn't butterflies. Still, I heard some wise words about monarchs before I left the Tri-Cities. The maids at the Vagabond Motel wanted to know what I was up to. Living out of a Honda Civic, I found that I needed now and then to empty everything and resort. Viewing my gear spread about the room, they couldn't imagine my business, and they weren't too shy to ask. When I explained that I was following monarch butterflies, Serena asked, "How much patience do you have? When I see a butterfly, I say, Bye-bye!"

"Yes," I laughed, "for me, it's mostly the same."

Pasco is a city of bridges. By early afternoon I had crossed all three meant for automobiles: a basic one, a handsome blue one, and a soaring suspension bridge with a single pair of central masts and a magnificent webwork of white cables. Cruising the east-side dike on the blue-collar side of the river, I found a series of old, dried-up "sanitary lagoons" at the south end of the railyards and truck docks. Sewage farms have long been revered as prime birding sites. I remember watching a stoat jump on the back of a moorhen in a sewage farm outside Cambridge, England. Tossed into the fetid muck by the tough, cootlike fowl, that was one disgusted stoat. Now I wondered if such places, dried out and gone fallow, can be good for butterflies, too, which frequently visit scat and manure for the nutrients they provide.

The pits were full of fresh milkweed and blooming Canada thistle and goldenrod. Right off, I spied a bright monarch over goldenrod on the raised border of the field. I waited too long to swing at it and was obliged to track it down into the pit. Once tagged and released on a thistle head, she waited a moment, then flew off strong and high, direct to the southwest. After a couple of football fields, she glided down toward the Snake River. Rounding the former lagoon I saw another monarch, standing out like an orange billboard among the lavender thistles. We repeated the same dance: the netting, the tagging, the release onto thistle; a two-minute pause, then the flight up and away for two hundred yards and the descent, but this time southeast.

Number 83 got a pass from a dragonfly on the wing; no. 84 got the once-over from a barn swallow, twice. The first time the swallow looked and veered. When it came back closer, the monarch avoided it by agilely dropping, very much as I've observed moths behave under bat attack. Then, falling toward the river, she crossed the glide path of a white pelican. Back in the pit, whites, sulphurs, and painted ladies lingered over the sweet-smelling thistles on this sour ground where shit had become first flowers, then butterflies.

At three I arrived at Sacajawea State Park and the confluence of two of the great western rivers, the Snake and the Columbia. I grew up, and first learned how closely butterflies are wedded to watercourses, along an irrigation ditch on the high plains of Colorado. In the hyperbolic language of its Victorian builders, a simple forking of the High Line Canal was called the Great Bifurcation. Here was a truly *great* bifurcation, though since it is where the rivers merge, it could equally well be called the Great Convocation of Northwest Waters.

I had pictured myself following the Snake out of the state in the southeast. But the river is hard to reach over most of its run to Idaho, and its upstream course actually curves north before heading east and finally south, where it emerges from Hell's Canyon. It was hard to imagine a migrant getting to this point from, say, Hanford Reach, then turning back north just so it could eventually travel south again. But I could imagine them flying southeast from here between the Blue and Wallowa Mountains to hit the Snake later, or continuing on down the Columbia and eventually abandoning it to fly south. Not that they all necessarily go the same way. Butterflies are *wedded* to watercourses, not welded to them. They employ rivers when rivers work for them, abandon them when they do not. The challenge now was to decide which river worked best: the Snake, the Columbia, or some overland average between the two. But I didn't want to figure it out; I wanted to be shown.

In the muggy afternoon, I made a river crossing on foot over the railroad bridge that no. 83 had flown toward from the sewage lagoon. Snake River Bridge no. 2 was a rusty old drawbridge that looked to be up for good. The tracks appeared unused. I strode out to the middle, where the immense counterweight reposed overhead. Scoping a similar bridge downstream on the Columbia I could see that the bridge tender had to climb sixty feet of caged ladder and sit in a hot tin box on top, above the raised section. Such a lot of iron to shift about at the whim of river commerce. It was spooky being out here — the river so big, so far below, the massive stone supports bereft of ladders. No migrants plied the river — just the motor barge *Prospector* from Vancouver, Washington, an immense, rumbling green slab, slightly raked. On shore, oil tanks and grain docks; rabbitbrush and asters beside the oxidizing tracks; bulrush banks and red osier dogwood, in unseasonable bloom. Cottontails and redtails, trying to get together or not, depending on their points of view. Someone shot at doves. Nothing stirred but me on my high perch.

Once back in the park I stepped onto a big navigational aid on the point of the peninsula between the rivers. Day-Glo pink and chartreuse, with a big letter S in green and a green light on top, the beacon stood on a concrete, beam-bound platform littered with goose poop full of little striped rugby balls — Russian olive seeds — and otter spraints full of fishy, clammy, crawdaddy bits. Geese cackled, magpies purled. From up on the otter's platform, I could see any monarchs that might be coming down either river.

Instead I saw the river tug *Sundial* motoring down the Columbia pushing a grain barge, its wake plashing onto the riprap by a bit of milkweed as a train whistled between the two railroad bridges. Five barges stacked with wood chips, grain, and containers grumbled down the Snake, shoved by one big Tidewater tug out of Portland, and made a tortured turn into the Columbia. And then, even as the tug lumbered beneath the downstream bridge, ye gods! the train rumbled over MY bridge. The horn blew, the counterweights swung, the center span of the bridge rose once more. It still worked after all. If I had still been out there, it would have been very awkward.

And then something truly amazing happens. As I walk around the point in this park at the mouth of the Snake River on this clear, sparkling, warm afternoon, an inviting stonework summerhouse lures me with its shade. Leaning against its low wall, my palms out on the cool basalt blocks, I ponder the courses of nos. 83 and 84 after they dropped toward the river. I say to myself, "I need to see a monarch come beating its way up or down the river through here."

As the thought settles, a monarch comes beating its way *down* the broad Snake, some twenty yards offshore. It wears no tag. I watch it fly right out of the Snake River mouth and over the Columbia. Then, flapping *very* close to the water, just inches from the barge wakes, it crosses the half-mile of open river before my eyes. It has nearly made landfall on the far shore at Two Rivers Park before I lose it in the complex pattern of wave-lap reflections. That amazing monarch is flying strong, pointing me down the Columbia — to the sun! I have my answer.

A fingernail moon came up, and my hands went pink in the lowering sun. Two white pelicans passed the moon, crossed the river, hit the sun, ignited, merged, separated, fizzled in the purple haze so only their cinder-black wingtips remained. Sixteen flaps and then a glide. I followed them for a mile as the sun slopped into an ocher pool. Some big fish splashed, probably not salmon, though the sky and the Columbia were the color of salmon flesh with blue streaks, as if taking on the substance of the fish that were no longer there. Here, after all, is where the nearly extinct Snake River salmon meet their bifurcation, too.

It seemed to me that Sacajawea Park should be a "power point" for monarchs, like Point Pelee on Lake Erie, Lighthouse Point in New Haven and in Santa Cruz, and Cape May in New Jersey. And so it was. Once the Shoshone woman got Bill and Meriwether to this point, the river did most of the rest, because they were bound for the Pacific. It was just the opposite for me. The river route had been clear so far, but the river must soon be left behind. Unlike the salmon, river-running monarchs travel cross-country as well. We would both need to find another route south.

I spent the night under sharp stars. A brisk, cool wind brought the morning, sunshine, and a monarch flying or blowing out of a locust next to my campsite on the Walla Walla River. The only others in Madame Dorian Park, a rough Army Corps of Engineers facility, were a pair of old hippies settled in for the full fourteen-day limit. Next to their trailer they'd laid down an Astroturf dooryard and set up a wrought-iron stand and hanger for bird feeders. I was just wondering whether monarchs use the invasive alien star thistle, when a pale one came off a clump of it and whirled around to a Russian olive — and I saw I had been fooled again. It was a big female viceroy, three inches plus. She basked, slowly fanning in the wind-blown sun, an orange advertisement of coming autumn, brightly backlit, spread in full view of birds that paid her no notice. Then a real monarch popped off the rabbitbrush between the highway and the old road. After I missed my wing shot on the stiff breeze, the monarch let the wind sweep it easterly over the yellow rabbitbrush plateau. Watching it go, I could easily imagine, as Robert Frost did, that "those great careless wings" were made "for the pleasure of the wind."

I followed, but it was all birds. Unhunted dowitchers safely grazed, and fifteen killdeer cut past a backwater slough, making their sweet racket. Over the water, barn swallows chattered their impatience to leave. A female northern harrier, honey-brown and white-rumped, hunted the flats. I drove east along the Walla Walla for a few miles. Along the edges of the river road bloomed Russian thistle — pretty up close, like a million little pink hibiscuses; but *Salsola* sticks to your skin and later grows into tumbleweeds, which stick to fences and culverts and the wind itself. A fellow drove up and scanned the river, probably a working man on a break from the nearby Georgia Pacific pulp mill, wishing he were fishing. We nodded but did not ask each other's business. The riparian border was a weedy, stickery jungle bounded by cottonwoods and willows. Forging through a floodplain summer-baked into six-foot blocks of clay with three-inch cracks, I entered a dense tangle of goldenrod bound together by ten-foot nettle stalks. It was so thick I had real trouble getting out again, and nothing flashed orange for my trouble.

I filled my water bottle in a mini-mart at the junction of the Columbia highway and the road to Walla Walla. Between green neon beer signs depicting a jumping fish and a cactus, a root beer readerboard announced: "Sturgeon Candy Engelhart 65″ 7-7-96." I think it meant that Candy had caught a big fish, and that it, not she, was five foot five. As much as I love fresh sturgeon, I would no more eat the bottom-feeding beast so near downstream from both Hanford and the paper mill than I would willingly make a meal of monarchs.

As it turns into the gorge, the Columbia River rounds its Great Bend and wends west through Wallula Gap, a major choke point for the floods from Glacial Lake Missoula. At peak postglacial flow, Wallula discharged close to fifty times the amount of water released by the Mississippi in high flood. The Two Sisters stand in the breach, dramatic pillars of Frenchman Springs basalt graffitized with "96" and peace signs. Having been two of Coyote's three wives, the sisters were turned into rock by the jealous trickster. I put up an osprey with a big fish in its talons. It spiraled off upstream for miles.

For most of its westerly run, the Columbia serves as the state line between Washington and Oregon. East of the Great Bend the border is just a straight line on a map. As I rounded the gap, I crossed that line into Oregon. Masses of rabbitbrush and sunflowers, suitable fodder for my quarry but sharply wind-tossed, stained the roadsides pee-yellow where the river became the frontier. But the next monarch I saw was miles downstream, just above McNary Dam.

In the late overcast afternoon I set out on the McNary Beach Trail to stretch my road-rubber legs. I'd gone no more than fifty yards when a monarch flew up over the bank and dove directly into a Russian olive to roost, too high to net. But instead of settling right in below her until dark and getting a solid fix on her final clinging position, I continued my walk east along the huge, flaccid pool of the

Columbia on the former railroad bed, hoping for others. A few minutes later I saw a cabbage white go to roost some twenty feet up in another Russian olive. The air was still now, but both butterflies had chosen the eastern sides of the trees against the morning westerlies. Returning to the monarch's roost tree at 5:30, I couldn't find the butterfly. It must have crawled farther in. But afraid that it had left, disturbed by a jogger, I tossed small stones. One actually struck the right limb, and the monarch fluttered around from the southeast side of the tree to the northeast. I couldn't spot it again and decided I'd just have to try to see it leave in the morning.

I bedded down in the car in the parking lot a few yards from where the monarch was sleeping. Downstream, McNary Dam and locks hummed. A big paddle wheeler joined a trim tour boat in the lock, both of them lit up like a carnival, for passage down to the next pool. The green starboard lights of a big ship slipped past in the dark. It was good to be there, with a fine, hoppy Grant's Scottish Ale to drink, Steinbeck to read, the river, and a monarch roosting nearby. When I turned out the light, I tried to compress my being into the butterfly's poppyseed brain and imagine what it was like to be clinging to a leaf in the dark, with a hundred miles to fly against tomorrow's wind.

First thing in the morning, louder than the trains and barges in the night, louder than the turbines of the dam itself, even louder than a jetski, was the industrial-strength riding leaf blower driven around and around the parking lot by an industrial-sized worker. I got up and went to stand knee-deep in the Columbia River, the water about the same temp as the air, a little below 60. My tree-watching vigil lasted all morning. The chief entertainment was a pine siskin trying over and over to get a drink. Each time it hopped to the riverside, a wavelet struck, causing it to fly a few feet back. Working along the gravel shore, the yellow-barred, brown-streaked finchlet did its little dance right past my feet. I had to laugh, though the poor siskin seemed doomed to go thirsty.

I also watched a furry tiger moth caterpillar walk right into the Columbia River. This got me thinking of the adaptive value of random dispersal. Rather than a sharp unidirectional pulse, the migrations of monarchs and many other animals actually describe a shotgun pattern, with much of the population getting to the most desirable destinations while some others go wildly astray (like the monarchs that arrive annually in Britain). For the northward migration this makes particular sense, since the milkweed resource is widespread. For the southward, this pattern might assure that the migrants spread across the full spectrum of acceptable winter quarters. Many other species, not necessarily migratory, spread out in all directions, seemingly to optimize their survival chances overall. So it is with woolly bears. By either strategy, some will necessarily drown in the process, or be crushed in the road, sacrificed to the experiment. I was going to save the damp larva, but a sneaker wave got it. A little after noon, in good full sun with slight haze, I decided my monarch had probably slipped out the back door, and I did the same.

From the mouth of the Umatilla River west, monarchs were moving along the Columbia shores. In a wildlife area east of Irrigon, Oregon, a swale between the roadside olives and riverside willows was chromium with rabbitbrush, fleecy with milkweed puffs on the wind. Checking some chewed-up milkweed, I spotted a mint-fresh monarch hanging from it. In fact, the small male had only recently eclosed. Nearby, the chrysalis shell hung colorless by its cremaster, hooked into a button of silk on the midvein of an unchewed leaf. I was not sure I should tag him, as his wings were still soft, but I managed it with great care. Placed on a Russian olive bough just above his birthplace, he walked around a bit, then lined up cleverly on a twig so that the westerly breeze would not bend his petal-soft copper wings, which felt like silk to my fingers.

I prowled the rabbitbrush upswale, where monarchs were attending to nectar. As the day waned, I revisited the soft, silky one, and his wings seemed firm and fine. Another monarch flew in from the east, swung around an olive two times, and alighted on the northwest side about twelve feet up. I considered camping here between the two monarchs, tagging the new one when it grew cold, and watching them both in the morning. I bent to the task of making a stick marker beneath the untagged one, looked up, and found it gone. Big skeins of ducks sutured the sky, on the move. I declined to remain after all.

So far I'd been obliged to follow the south (Oregon) shore, since no road continued down the river on the Washington side from where my Sacajawea had crossed the river before my eyes. Now I crossed over the Umatilla Bridge back into Washington, continued downstream, and camped to the lights and whistles of freight trains at Crow Butte, almost an island in the reservoir behind John Day Dam. Morning came on a cold wind that pinned a pale monarch to the marsh in the crook of the island causeway. Far downstream and a few years before, a fisherman friend named Dan Pentilla had introduced me to a term for a concentration of whatever one happened to be hunting. When we found a spot in the river where the silver salmon schooled, Dan called it a fish patch. Now when I implored the butterfly gods for a mess of monarchs, I decided "fish patch" was as good a term as any. Today, the last day of summer, the gods were good to me.

Mount Hood loomed off to the southwest, and beyond it lay the edge of the world as far as monarchs were concerned, since milkweed promptly drops out farther west. The highway traced a low shoreline between dust-dry hills, punctuated by little bars bearing milkweed, locust, and licorice. The first trees for many miles appeared in a thick grove at a place called North Roosevelt. The trees, tall elms and locusts, defined a horseshoe embracing a bunchgrass mini-prairie. Within the tree ring, twenty-four basalt boulders stood on concrete plinths, ranging in size from a microwave to a small fridge. Each boulder bore Indian petroglyphs — inscribed relief patterns of lizard, bighorn, and other motifs. Most had colors leftover from pastel rubbings people had done; others had been painted over. At the base of one, letters crudely inscribed in concrete read "East Klickitat Gem & Mineral Club, 1964." I rightly surmised that the club had transferred the rock art from cliffs along the Columbia to save them from flooding by the John Day Dam.

Between the park and the highway ran a low swale of burdock and milkweed. Checking it out, I saw one, then two, monarchs fiddling around their host plant. A formidable fence stood between us, and when I got to them they sailed over a hedgerow of Russian olives. This play went on for quite a while, as I played their fool repeatedly. Then I spied a *very* tatty female nectaring on a four-foot goldenrod. I netted her, but she got out when the bag flipped over, and laboriously flew off across the entry road to the eastern hollow, a hundred yards away. A few minutes later, to my astonishment, I saw that she had returned to the same inflorescence of goldenrod. I carefully caught her this time, handled her gently, tagged her, replaced her, and she resumed nectaring without hesitation. She knew exactly where she wanted to be, and no one would keep her from it.

This venerable, ragged mother had only a third to a half of her wings left. Her body and wings were greasy. Also, some of her veins were squiggly, showing developmental difficulties. I softly palped her abdomen but could feel no spermatophore — normally, an older female would have one or several of these seed packets deposited by the male, unless she was of the migratory generation that would not mate until spring. She might have been old enough to have absorbed any spermatophores she once possessed. It was remotely possible that she was a survivor from the spring's northbound immigration, or even, conceivably, from the previous autumn's emigration. If so, she would be a full year old, though this is unlikely; I have heard of only one monarch living a year, and very few of the fall departers are thought to make it back north the next spring. She was still tough, and it is possible that she belonged to a recent generation, was no more than five or six weeks old, and had simply lived hard. She wasn't telling. But no matter her age, she taught me the valuable fact that a monarch can learn, leave, and regain the location of a particular nectar plant. I had not known that.

Soon I saw a big male and caught him. Quite possibly a son of the sedentary old gal, he had only recently eclosed nearby. After being adorned, he perched on an olive, spread, quivered, flicked his tag-wing two or three times, and lifted off to the west. A gust drove him up and back, and he dropped down into the brush. When I looked for him, I fell into a treacherous maze of deep badger diggings and tried to break an ankle. And there, my face near the ground, I found Washington's other species of milkweed: *Asclepias fascicularis,* narrow-leaved milkweed. Its thin, willowy leaves make it much harder to spot than showy milkweed, which was here too, along with half a dozen monarch larvae, feeding side by side on both species, just a few feet apart.

There were others, and finally one big monarch floated over the swale, evading me, eventually alighting in a tree in the center of the rock horseshoe, then rising again and drifting away. It was the last I would see in this veritable swarm, the largest one-time concentration so far on the expedition, quite justifiably known ever after as the fish patch.

After Roosevelt, everything changed. Hackberry trees appeared, then Oregon oaks. Showy milkweed dropped out in favor of narrow-leaved (also called Mexican whorled) milkweed, which I learned to recognize by its slender, bursting pods. It wasn't very far west to where the Cascade crest runs down to the river at the Skamania County line, the maritime green and Douglas-fir reassert themselves, and milkweed disappears. Very soon the low hills over in Oregon rose up to form the high-cliffed, waterfall-laced canyon for which the Columbia Gorge National Scenic Area was designated. I knew that monarchs must soon strike south: they *have* to cross the Columbia. When I finally witnessed such a crossing, it was in one of the strangest places in the Pacific Northwest.

The railroad maintenance road under the cliffs beyond John Day Dam was rocky and slow but passable all the way to the apricot and peach orchards and whitewashed church steeple of Maryhill. Sam Hill, a remarkable public figure, intended Maryhill (named for his daughter) to be a model Quaker farming community. Above, perched on cliffs like a Rhineland fortress, stands the stately home he built for his wife, daughter of the (unrelated) railroad magnate James Hill. When she wouldn't come to the windy, lonely mansion, Hill invited the exiled Queen Sophia of Romania; Alma Spreckles, the sugar heiress; and Loie Fuller, the popular dancer and film star, to visit Maryhill House. They helped to assemble the important collections, including Rodin sketches and sculptures, that make Maryhill a superb art museum today. Hill began the building of the revolutionary, reinforced concrete mansion in 1914, but it was not completed until 1940, well after his death.

Maryhill also has splendid gardens, shielded from the ever-present gorge gales by high windbreaks of trees. There I found monarchs dropping from the sky to take the nectar of purple and yellow butterfly bushes, pink cosmos, lavender asters, and blue salvias. It was easy to catch and tag them as long as the sun lasted. Each time I finished one, a new one appeared. If the Roosevelt swale was a fish patch, this was like shooting fish in a barrel compared to the chases I was used to. One had a broken costa, the stiff leading-edge vein that supports the wing like the strut of a kite, and I was able to splint it successfully with the tag so that it could lift off again.

Perhaps they had come on high from the Yakima country to the north or had worked their way along the Columbia. Some had arisen on the premises no doubt. Those that left sailed south. The mansion blocked my view, so before the day grew too old, I dropped down to the rabbitbrush bench just above the Columbia shore. There I found still more monarchs, harder to catch, visiting the wild plants. As I released a female, she fulfilled my urgent hope and headed out over the open river. I followed her with my binoculars as she danced with the whitecaps, brighter and more adept than any of the colorful windsurfers that throng the river. The channel narrowed here to just half a mile, and she aimed southeast, directly at the grain elevator of Biggs, Oregon.

Above and a little east of Maryhill stands Sam Hill's memorial to peace, a concrete replica of Stonehenge, now an eerie echo of the Roosevelt rocks. I had camped there one summer, on the very lip of the basalt flow, and had awakened to the boundless river and gorge far below. Now, sunset was near, and I considered bedding down there again. But I made a different decision. I crossed the bridge to Biggs, on the track of that last monarch.

SIX

Deschutes–John Day

In the bright morning at the mouth of the Deschutes River, steelheaders and mergansers fished. The water reflected its far shore, burnt black from an errant railroad welding spark. A train tunnel emerged from the basalt face, spitting slow freights.

The monarch that had led me across the Columbia made landfall in Oregon at Biggs Junction. The way south from there rises to a plateau between the Deschutes River on the west and the John Day River on the east. Both of these big tributaries of the Columbia wend northward out of Oregon's interior highlands, like two snakes trying to slither in parallel. Monarchs are known to frequent both watercourses.

I had with me atlases of all recorded Northwest monarch occurrences, compiled by John Hinchliff. This meticulous British-American architect, living in Portland, served the region's lepidopterists as scribe for their collective data. While agencies in both Oregon and Washington fiddled around, year after year, trying to establish digitized data bases for their faunas, John proceeded to manually map and publish the accumulated results of a century of butterfly studies in the two states. I reviewed the monarch maps dot by dot in search of clues to the monarch's possible route south.

The Hinchliff atlas showed monarch records at the mouths of both rivers and here and there along their squiggly routes. It has long been assumed that monarchs this far west winter along the California coast. If they are going to avoid volcano after Cascade volcano, with their associated harsh weather, and find nectar as well, the butterflies would profitably work these rivers southward until they broke out into the basin-and-lake district on either side of the California border, then hop over the Cascade/Sierra gap toward the Central Valley and, ultimately, the coast. So I decided to work south along the rivers myself, as best I could.

No road runs directly south from the mouth of either the Deschutes or the John Day. To approach the Deschutes around the side, I first drove above the Columbia Gorge in the direction of Celilo Village. A western meadowlark fluted over a vernal swale, its yellow breast reflecting solid acres of star thistle. Celilo is where the Columbia River had its greatest rapids and the native people did their best fishing. It is now a prefab village above the reservoir beside the freeway, the home of those who remained after The Dalles Dam did its dirty work of drowning the falls, wrecking the fishery, and killing what was left of the river culture in 1956. Salmon's revenge for the dams is the immense headaches that salmon management has become today. The butterflies' payback for the rich, flooded riparian habitat is the endless nectar supply brought by the star thistle, knapweed, and other alien weeds that render the rangeland above the water level useless. The Indians' inadequate but costly revenge may lie in reservation casinos where the hunter has become the hunted.

In this stony land, people make use of rocks. Across the Columbia from Celilo I could see Wishram, a funky jumble of railroad town and hobo junction. On its waterfront stands a monument, twin pillars of natural columnar basalt bound with cast brass ropes, dedicated to the pioneers who came and claimed the Columbia. Farther west lurks She Who Watches, or Tsagagalal, the greatest of the Columbia petroglyphs, whose image has been copyrighted as a trademark by the Columbia Gorge Interpretive Center in a chilling act of cultural appropriation. But the dams are the most daunting rocks of all. Or are they? The river will have its way with them in the end, as it does with all rocks, with the very lava flows themselves. The ten-foot face of She Who Watches may see Celilo Falls come back some day.

I was going to have to reach the Deschutes corridor by driving south out of the city called The Dalles, so once more I touched down on the Columbia's shore. I ended up spending the whole afternoon walking the riverside trail in the languorous sun of the first day of autumn. What seemed like ten thousand yellow-rumped warblers and the odd fishing osprey occupied quiet coves where bright pink mallows studded flood-flats. Every butterfly one could expect abounded, except the one I wanted. I did find a raccoon's tail, already skinned out and cured. I took it with me, but didn't tie it to Powdermilk's antenna nor to my hat, though Davy Crockett had been a major obsession of mine about 1955.

Dark fell in The Dalles. I nearly ran into Preacher Rock, yet another stone monument, which echoes the one in Wishram but stands in the middle of a street. After a long search up and down vaguely familiar streets and bridges over identical ravines, I found the little ten-buck cabins surrounded by violets where Thea and I had shared a courting tryst one wildflower spring gone by. I'd hoped to sleep there again, but the cabins were boarded up and falling in. As I cruised the dim streets of The Dalles looking for a cheap but decent meal, dimmed a little myself by the inexorability of change, I meditated on the word *funk*.

I had called old Wishram *funky*, an adjective by which we mean ungentrified, maybe seedy or shabby, but interesting, with character. Then there was the *funk* I had worn last week, and would acquire again, when seriously in need of a shower. And the *blue funk* I felt when I finally capitulated to hunger at Burgerville, tired and a little lonely, not real damn sure what I was doing or where I would sleep, and too keenly aware that home lay just four hours downriver. On my tape player I listened to Rod Stewart's "Every Picture Tells a Story," with its punning line "My body stunk but I kept my funk." And when I switched over to the radio, I swear I heard George Clinton of Parliament Funkadelic (aka P-Funk), on a blues program, asked to define the term's musical meaning: "Blues with a groove, blues with a beat," he said. That cheered me up. I decided to sleep on it, and look it up in my Funk & Wagnall's when I got home.

* * *

Highway 197 south traversed a tawny topography, suddenly autumn. Beyond Dufur, Shady Brook Road led to rural Tygh Valley, where milkweed picked up. Getting out to examine it, I fixed on the vivid green of a monarch chrysalis. But it was on the wrong side of the leaf — a fat Pacific tree frog hunkering in the leaf's curl. The heads of slender horsetails growing among the showy milkweed looked very much like young monarch larvae. But these stands bore none of the guild of major milkweed insects — bugs, beetles, spiders, or butterflies. Maybe it was the edge-of-range effect showing, whereby many species become uncommon as they approach the geographical limits of their existence. A coyote stared and trotted beside me as I sped up again, into juniper.

Impressive, several-tongued White River Falls embraced a turn-of-the-century hydropower works. Its gears, pipes, pools, dams, chutes, races, concrete works, masonry, and buildings, all in disrepair, clogged a dramatic basalt gash in the earth. A canyon wren sang its cascading tremolo, as happy with concrete as with sandstone, and rock doves played, as much at home as on their ancestral Atlantic cliffs, watching for the peregrine. Violet-green swallows massed for their remove to the tropics, as one of the four redtails circling about sifted down through them. I'd seen buteos pick off bats this way, but the swallows pestered the hopeful hawk on its way.

The day had gone chilly, an impression perhaps reinforced by the name of the classical music deejay I found on a farm college station: Gillian Coldsnow. Boccherini drifted into farm reports as I left the highland and dropped down Winter Water Creek to Sherar Bridge, where I joined the Deschutes River at last. The last time I'd been here, blue-eyespotted Hunter's butterflies had been all over the thistles, and Indians on rickety platforms dipped salmon from the falls below the bridge. Today there were no butterflies, and the fishermen were old white guys on shore with fancy rods and reels.

An ancient Indian trail crossed the river here, and wagons were floated across until the first toll bridge came in 1860. The rails arrived in 1916, opening up this canyon. As was often the case, competing railroads shoved iron up opposite sides of the river. The line on the east bank failed after ten years and later became a rough, dead-end auto road. James Hill, father-in-law of Maryhill's Sam Hill, prevailed on the opposite bank with his Oregon Trunk Line. His successors still ran three or four trains a night, each of which I got to know intimately from my campsite, opposite the tracks at the Blue Hole.

At that deep eddy, the river ran broad and deliberate down to a rapids. There were no rafters here now, just one party of fishermen in a camper. As I dropped off, I listened to the tumbling Deschutes. Then I was lying in a raft, gazing up as orange monarchs drifted over, more and more of them, faster than the river's flow. I was immersed in a molten spume of shimmering butterflies. Each one resolved for an instant, floated like a reflection of the sun in its own eye, then melted into the golden whole. A warm satisfaction came with them, the sort of rich fulfillment found mostly in the best kinds of dreams, and childhood. When I awoke to more cold clouds, I wondered where all the butterflies had gone.

I needed sun. Hot oatmeal and coffee helped. Two kinds of *Rhus*, poison ivy and nonpoison sumac, lined the opposite shore. Both shone scarlet when blessed sun hit the canyon at 9:30. My plan was to motor north down the dead-end Deschutes road as far as I could go and watch for monarchs motoring upstream (south). I hit the road with Van Morrison singing "Autumn Song" and a fresh peach in hand. Beyond Oak Spring a pink penstemon bloomed above the road, and a pink monkeyflower, smaller than the alpine kind that springs from mossy banks of high rills, grew in the dry rocky verge: both beautiful, in lavish colors you don't expect in dry places in the autumn. The river had carved caves and plucked overhangs from the chunky basalt. Alkali-streaked pillars splayed like art deco sunsets from vernal waterfalls. The White River, miles below its falls, plunged beneath James Hill's bridge to its outfall. Female common mergansers fished and basked and preened their hennaed butch-cuts in the river mouth.

Humans mimic mergansers by rafting and fishing the popular Deschutes. Checking each of the Bureau of Land Management riverside access sites for nectar, I found them battered by rafters and fishers. Permits should be awarded based on per capita litter removal, including pounds of filter tips and yards of monofilament fish line hauled out. The government and tribes manage the river cooperatively. A Warm Springs Indian tribe takeout called Sandy Beach (a euphemism for basalt blocks) lay just above Sherar Falls, gray yesterday, turquoise-green today. An Indian in rubber boots, blue jeans, white T-shirt, red billed cap, and long black hair, cinched to a safety rope, was dipping from one of the platforms built out over the white cascades. These fishers use no rods, lines, or hooks, just huge nets with wooden pole handles that dwarf even Marsha. Boulders anchored the base of the platform, a pine log and slat made a seat. She put down her heavy net and walked over to talk with a handsome Warm Springs tribal fish cop in a truck on the road.

Watching the two of them talking, I was reminded of a conversation I'd had with three Yakama Indians in a café on the Columbia, after seeing them out on a boat early that morning. They told me they had seen not many monarchs on the river but lots on the Yakama Reservation, where milkweed proliferates. Then they asked if I'd noticed any other boats, because their nets had been cut. I told them of a white pickup that had come by, and they thought it might have been the fisheries patrol. Since the dams were built, the Indians have had a hard time transforming their treaty rights into a livelihood. With some whites thinking the Indians are allowed to take too many salmon, campaigns of intimidation are not unknown, and fish and game possession laws have been applied with a heavy hand. And, of course, the fish runs are nothing like they were in the past.

The woman dipping at Sherar Falls didn't seem to be having much luck. Farther on, I passed a sport fisherman's rig with the vanity plate "*Salmo g.*" Two men in a red canoe lolled on the opposite bank. Around a bend, a sleek little wooden steelhead drift boat floated at anchor, its three folks lunching in the shade. I wondered whether any of the people here today, some of whom had struggled up in motor homes the size of boxcars, were having any better luck catching *Salmo gairdneri* than I was with *Danaus plexippus.* I thought of the differences between our methods: casting baited hooks into the water and waiting for invisible fish to strike, versus baitless stalking of conspicuous prey that might or might not appear, both ways equally beholden to hope and luck. It seemed to me that the Warm Springs woman, netting the flashing salmon as they leapt from water into air, came as close to knowing where these disparate passions meet as anyone ever could.

Below Buck's Hollow, coronis fritillaries nectared at rabbitbrush. Ambush bugs mated nearby, and I wouldn't have been entirely sorry to find that they had snagged a monarch with their mantislike, lightning grasp. The road turned rough, the canyon grew high and layered and red. Big Beechey ground squirrels scampered as if bits of the speckled gray pebble road had come alive. An older couple in an Oldsmobile stopped, and the man asked me, "If a guy was to go down this road, would it come out, or would you have to drive on back?" By "come out," he meant could he reach I-84 at the river mouth. I told him he could drive another five miles or so, and then he'd have to double back. "Well, it's pretty," he said, "but not *that* pretty," and he turned around.

I disagreed. At Rattlesnake Canyon, in place of the advertised serpent, an osprey topped an incense cedar on an island, and in the top of a single dead tree perched a yellow-rumped warbler, an American robin, and a northern flicker, side by side and apparently indifferent to one another. Lemony sulphurs and glittering little acmon blues occupied the snow buckwheat. A glimpse of the blues' feathery white fringes, fiery orange lunules, and scintillating turquoise spots beneath, their river-blue wings when they spread to the morning sun, and I felt the road was worth every bump.

To the north, the road died in an impassable jumble of rock-slide rubble, broken bridges, and old hopes. There were supposed to be California bighorn sheep here, according to a BLM interpretive sign at the end of the road. It said that native bighorns appear in local petroglyphs, as I'd detected on the carved rocks at Roosevelt. The function of rock art is poorly understood today, but some think that depicting an animal was either a record of past hunts or a form of supplication for the animal's complicity in its own pursuit. Hunting was better in the old days for salmon and sheep, portraits or no. There might have been more monarchs, too; there surely couldn't have been any fewer.

I left the Deschutes and struck east for the John Day, hoping that river might prove more generous. The sun dropped, the moon rose as I approached the Deschutes/John Day watershed via Cottonwood Canyon. From the high flat between the rivers, Mount Adams was a purple hump to the northwest in a salmo sunset. Gazing up and down the Cascade spine, I could see Mounts Rainier, Adams, St. Helens, Hood, and Jefferson, all five peaks cast in pink and purple. The land dropped away again as I entered the drainage of the longest free-flowing river in the forty-eight states. Over 380 miles in length, the John Day was named for a mountain man who accompanied the Astor-Hunt Expedition of 1811–12, who stayed on to explore much of the country south of the Columbia. John Day's given Indian name was Mah Hah, and in his classic book *Astoria,* Washington Irving described Mah Hah as "six feet two inches high, straight as an Indian, with an elastic step as if he trod on springs."

I struck the John Day at J. S. Burnes State Park with just enough light left to have a look. The riverside was all private and posted to the north, so I walked a mile of shore on the southeast side of the bridge. The riparian fringe was woven of blue willows and tough little hackberry trees *(Celtis reticulata).* Hackberry butterflies, which feed solely on *Celtis* leaves as larvae, are not known north of Utah. Still, I scanned the undersides of the elmlike leaves for the remains of *Asterocampa* chrysalides, which when full are as green as monarch pupae, but thorny.

I leaned against the basalt, whose lava-bubble vesicles were full of life, like tiny planters stuffed with miniature cushion plants. The smooth, lichened wall felt warm from the day's thin sunshine. Cliff-swallow nests poked down like swollen downspouts. Monarchs could have roosted inside them, the red-rumped builders of the mud pots having long since preceded the butterflies south; but I saw only an ochre ringlet go to roost in the white-fluff seed heads of wild clematis known as old man's beard, hanging from a hackberry. Down on the riverbank, a remarkable variety of creatures had signed the well-trodden backshore sands. Deer, coyote, rabbit, beaver, raccoon, an array of rodents, weasel, river otter, quail, heron, crow, and killdeer had all walked here in recent hours. The hills around are all grazed, but no mark of cow showed along this shoreline. The panoply of tracks told of a dozen food chains and a wild river that still works.

Morning brought clarid vistas and a day that would grow hot. The next place I could get to the John Day took me across the route of the Oregon Trail. The hills were yellow with large and small kinds of rabbitbrush, a nectar pasture for aerial livestock. Chukar and quail skittered through hot, dry Rock Creek; orange and yellow sulphurs puddled at damp spots, clustered and flew by the hundreds over the yellow composites — how I wished they were monarchs! Shrikes on a wire, a parent and young, made me wonder whether inexperienced shrikes ever catch, impale, and store monarchs on barbed wire or thorns, only to return to a nasty surprise.

The John Day River here was a series of blue loops through ocher hills. When I arrived at a locust-shaded spot on the bank, I walked vigorously in both directions, checking nectar along the shore. The road ended upstream at a farm. There was no one in sight, so I stripped and swam, gradually immersing scrotum, belly, kidneys, nipples, shoulder blades in the *cold,* clear John Day. Smoothed, spalled-off columns of basalt lay in the water and along the bank like long loaves, but I drifted over a soft silt bottom. Buoyed by ice water out of the highlands, I felt as sleek as otter, stone, or salmon. But in the rivers of autumn, the line between sublime refreshment and hypothermia is a slender one, and eventually I was obliged to crawl out into the sun, my full land-weight restored disconcertingly. Back on dry land, scores of iridescent tiger beetles flew from my blue feet.

Downriver by another ranch, a sharp-shinned hawk was hunting smaller birds among blue willows. Mark, a young man from Arlington, drove up in a Plymouth. He'd caught a steelhead and shot a rattlesnake. "The landowners laid a dead one out in the road, to intimidate us," he said. "They tried to shut us out, but we sportsmen demanded access to the BLM land. Well, rattlesnakes always come in twos, so I was on the lookout. He was actually going away from me — I had to pry him out to shoot him."

I said nothing. Mark paused for a moment, rubbing his sweaty head. "So, really, I was the aggressor," he finally said. "My dad'd say, 'Whyn'tcha leave the snake alone?' I like rattlesnakes, I really do. But I don't want him bitin' my son, next time we're up here." Mark showed the snake to me — shot to bits. "I put it here by the A.C.," he said, "so it won't rot too fast." He didn't show me his fish. He said he and his wife like to skinny-dip just where I had, and offered me a shot of schnapps. When I declined with thanks (having just had a tall Hamm's), he took his snake and left.

Hamm's would not have been my beverage of choice in a state famous for its pioneering microbrews. In the small-town store, all the choices were bland American lagers. I remembered the beaver, bear, and squirrel ads ("From the Land of Sky-Blue Waters") from my childhood, and the price was right. It was almost as refreshing as the river dip on that sweaty afternoon, and about as flavorful. I'd foolishly failed to pick up a case of Full Sail Amber Ale, brewed in the Columbia Gorge, when I was in The Dalles; now it was likely to be Hamm's for days, if not weeks, to come.

A contrail slicing the near-full moon started purple, went pink, then white, as the sky shifted into lilac. I made Fossil by dead sunset. Wheeler County's fancy big three-story, red brick, 1901 courthouse, with one pyramidal tower, one octagonal, stood out in a town of small scale, which is the point of all county courthouses. The modest bungalow across the street stood out, too, with fifty-four kettles and teapots lined up along its garden edge. Not far out of town, I made fragrant camp in a high, piny place called Bear Hollow, where autumn was well along; it nearly froze that night.

The morning was still too cold for monarchs. They can fly in the fifties if they have been active or basking, but they don't usually start up until the temperature reaches 60 or so. Down the road a park donated by a pioneer rancher was full of fresh-faced girls on a school camp-out, escorted by their moms, all in bonnets. As if at the tinkling laughter of the girls, who were walking by twos, butterflies came alive in the form of a very bright ochre ringlet and a mourning cloak playing around a spring. The next park along was paid for by the Oregon lottery and bore hopeful signs asking visitors to please water the trees. Several of the state parks had posted signs saying they would soon be closed because of tax limitation measures passed by the voters. They won't pay taxes, but they gladly buy lottery tickets. More and more, essential services are funded by chance.

My next destination was federal, run with funds that can't be withheld by citizen taxophobes unless they happen to be in Congress. The town of Fossil takes its name from the extensive prehistoric deposits that gave rise to the John Day Fossil Beds National Monument. I particularly wanted to see the Painted Hills unit of the monument. With autumn advancing, I knew I should keep moving until I saw monarchs, so I took Highway 19 east along the river, but then I turned around and drove west to the Painted Hills after all. It was a beautiful drive past century-old farms in narrow valleys, over a pass dripping with nectar of thistle, rabbitbrush, sunflower, even milkweed — as good a place to find monarchs as any. In Mitchell a bunch of hunters in big rigs were waiting for the operator of the only gas station anywhere around to come back from a random break. Gas-sipping Powdermilk still had a quarter tank, so I drove on.

The Painted Hills are an erotic Neapolitan pastiche of pates, breasts, buns, and knees sculpted from strawberry, pistachio, and French vanilla scoops. The vanilla is composed of ash/claystones of the upper John Day Formation, 20 to 22 million years old; the pistachio is ignimbrite of the middle John Day Formation, 22 to 28 million years old; and the strawberry, 28 to 35 million, from the Lower JDF, is bentonite rich in iron oxides. Fossils of camels, oreodonts, rhinos, and giant swine stud the beds, but the surface is fragile and you can't poke about at will. I took the boardwalk around one pink bosom and tracked down several solitary vireos singing from junipers all around Leaf Fossil Hill. Dawn redwoods *(Metasequoia)* once grew here, but now plants are nearly absent from the nubbly clay; even the tough junipers grow chiefly in the creases between them.

An old cowboy with his black border collie, herding cows just below Painted Cove, gave a hearty "Hello, how are you *today*?"

Since the Painted Hills are National Park Service land, I had not expected to see cattle there. Later I asked a ranger about the cowboy and cows. "Oh, that's Corky," he said. "We got this land from him, and he agreed to keep the cows below the county road. We *got* to do that." By making such an agreement, NPS managed to keep the compacting bovines off the delicate weathered ash formations, and Corky still had his grass, for now. Cattle have shaped this land and flora, and they won't go away tomorrow.

The ranger said he had seen monarchs over the Painted Hills, but not yet this year. "I did see bluebirds migrating through the other day, though," he said, "one after the other, right across the road." He recalled that when he was stationed on Alcatraz, monarchs came right over the island each fall. As a ranger myself long ago, I watched monarchs leaving the meadows of Sequoia National Park in late summer, sailing down out of the Sierra, perhaps heading for Alcatraz.

To avoid backtracking, I drove the Twickenham Loop, a long dirt washboard, but full of nectar, painted ash layers like lumpy-knuckled lion's paw scallops, and great rocks. It took me by the Monk's Hole, an enchanting spot on the John Day with diving rocks and basking spots overhanging blue pools. I met an old couple poking about on a four-wheeler, camped at leisure, virtually alone in this big country, and I could see Thea and me in twenty years, maybe minus the all-terrain vehicle.

Completing a long circle at Muleshoe Bend farther up the John Day River, I watched three species of butterflies bed down. As I was inspecting a small stand of milkweed, the corner of my eye saw an orange-flashing creature fly into a big ponderosa pine. At first I thought it must be a small monarch, but then I watched two red admirals go into the same tree. One went to roost where two pine cones joined, twenty feet up, while the other came back out to bask on a sandbar willow. So the first was probably *Vanessa atalanta* too, no matter how much I wanted it to be a wanderer. A woodland skipper and a cabbage white selected separate hackberries for their night's lodging.

This was the night of a total lunar eclipse, and maddeningly, as the hour approached, I found myself trapped in canyons. I drove like hell, but each time I thought the topography was about to roll out flat to the east, a curve brought on a new blocking ridge. Where the three forks of the John Day River diverge, dark came and I was racing the moon. I flew like a hasty bat down Picture Gorge, and emerged to a valley view of Luna one minute before totality.

This was to be the only total eclipse of a harvest moon for fifty-five years. The hills here were still painted dusty rose, and the disk of the moon at totality became the same color, as if it had only recently been carved out of the earth and had levitated a few degrees. As I watched, I drank a Hamm's, ate a Hershey bar, and listened to John Updike reading something sensual on tape. Finally I sat on Powdermilk's hood in the chill postdusk and listened to the coyotes keening at the rosy ghost until farm hounds drowned them out.

Afterward I set up my Gaz stove on the concrete base of a historic sign, boiled water, and celebrated the moon's return with Cha Cha Chili and hot spiced cider. A flight of bats or birds crossed the lunar face at the moment of full return, like a welcome-back party. The first time I went lunar birdwatching, we trained a spotting scope on the full moon, watched, and waited for migrant birds to cross over. When a perfectly recognizable heron did so, I was amazed and entranced. Tonight's eclipsed moon had worn the hue it might have had if a flight of monarchs had passed, trailing a haze of ruddy dust. Now I imagined the scales drifting away as the moon returned to cream.

The next day, at the national monument visitor center in an old homestead, the park rangers knew plenty of history and geology but nothing of migrating butterflies. These beds, where vertebrate relics are abundant, yield no fossil butterflies, though nearly modern species (nothing like monarchs) are preserved in ashen layers from Miocene lakebeds in Colorado. In chilly, windy sun I saw Picture Gorge, which had been a black blur the night before. Pictographs — painted images, as opposed to the etched petroglyphs of Roosevelt — adorn the 16-million-year-old walls. "It is said," says a brochure, "that centuries-old Native American trade routes between Canada, Mexico, and the Great Plains intersected here." Doesn't it stand to reason that centuries-old butterfly routes might pass this way too?

I merged into speedy State Highway 26, designated as a Journey Through Time": "This tour route celebrates an area uniquely rich in history, a place where fossils lie abundant on the landscape, and the Old West happened just yesterday." But 26 was being widened through historic Dayville. The espresso woman was distraught, having just learned that the project would take her graceful old cottonwoods. So much for history.

My monarch trail had definitely cooled. I felt like a sailor without a chart. I called friends and butterfly watchers Sue and Jim Anderson in Sisters, near Bend, way up the Deschutes from where I'd left it. Sue told me she had recently tagged twenty-five at Summer Lake and Tule Lake and easily could have doubled that number. I called home and learned that David Myers, a butterfly photographer and neighbor of ours, had recently found monarchs working rabbitbrush between the town of John Day and Burns, due south of where I was calling from. I was tempted to turn south toward Malheur National Wildlife Refuge, then southwest to Sue's productive tagging grounds along the California border.

But the fact that monarchs bred throughout that country did not mean they were in any way related to the ones I had been following from the north. Linking up with them as they departed, likely for the coast, would be the easy way to rejoin monarchs. Thea, though, had wisely cautioned me against moving the Ouija board myself. I needed to put my itinerary back in the hands of my guides, the monarchs. And that last one I had watched, crossing the Columbia, was bearing southeast, as had many of the others I'd seen. Maintained, that path would have taken it toward the Snake River and Idaho. In the past I had watched monarchs moving along the Snake. The following year I would find a good many on the South Fork of the John Day — but now I did not, so I struck a course toward the Snake.

Milkweed increased to the east, but the traffic prevented caterpillar stops. "I Brake for Larvae" would have gotten me rear-ended. Spotting a chance, I pulled off on a farm track and flagrantly trespassed on a posted bank above the John Day, now diminished in size to a rural stream as I neared its headwaters. The bank brimmed with almost all of my autumn butterfly friends, including bright pairs of purplish coppers *in copulo,* trying to get off yet another generation before the freeze, and two species I hadn't encountered before on this trip — the zephyr anglewing and Milbert's tortoiseshell. Both of these are bright enough, with their tawny and orange-peel banded uppersides, to give a start to an overeager monarch hunter. Both are also bark-dull below, and, like mourning cloaks, they hibernate for the winter as adults, taking shelter in a hollow tree. Monarchs lack the ability to pass the winter in metabolic diapause, as these species do — leading to the evolution of migration as a spectacular alternative.

But these overwintering butterflies also migrate, if not so far. They move, generally, northward and uphill in the spring and southward and downhill in the fall. In that way they utilize the full available range while occupying the most favorable regions for survival in winter. The migration pattern of red admirals lies somewhere in between this one and that of the monarchs. They infiltrate the North in spring but are not as freeze-hardy as the tortoiseshells. You can see them traveling en masse in the autumn toward warmer climes, but they apparently don't perform a long back-and-forth as the monarchs do. Beyond Prairie City, all these butterflies were moving across Dixie Summit, between Blue and Strawberry Mountains. Admirables, mourning cloaks, and Milbert's tortoiseshells crossed the road before me again and again. Watching in vain for a monarch among them, I began to wonder whether I should be tagging these butterflies instead.

When I had gased up in the town named for John Day, I'd cleaned two days' worth of insects off the windshield. Now, *smoosh!* — a beautiful big mourning cloak spilled nectar and guts all over the glass. *Nymphalis antiopa* is one tough butterfly, but I hadn't known how tough: scraped off, its abdomen macerated, it flew over the road and into a cottonwood, and perched. This gave the lie to a common cliché, that all butterflies are fragile. Anything that can live through an Alaskan winter out of doors, as mourning cloaks can, or carry itself from Manitoba to Mexico, as monarchs do, is patently not *fragile.*

I put Powdermilk up a punishingly steep and rocky road to Dixie

Butte Lookout, 7,592 feet above sea level. From firs and purple-scarlet huckleberries we came into sagebrush running up the sunny south side of the subalpine, with gnarly, ancient whitebark pines on the north side, and a gnarlier shrub with leaves like tough little knives in between. A sandwich for me, a pint of Quaker State 10/40 for Powdermilk, and a look about the windy top for hilltopping butterflies. Hilltopping is a courtship technique that concentrates many species (but not monarchs) on promontories, where males find and quickly mate with females. Checkerspots, swallowtails, and skippers often abound on summits. Here, too late into autumn, there were none. Nor was anyone home in the occupied and furnished fire lookout, which enjoys a fabulous 360-degree view into the Strawberries, the Blues, the Wallowas, the Ochocos, even the high Cascades. I could see why various writers and lepidopterists have cherished their jobs tending lookouts.

I had brought no camera, not wanting to be constricted by a 35mm view of the horizon. But here is a snapshot I wish I had: Powdermilk perched on that sagey-piny summit, the track to the sage green fire lookout, all against a brilliant blue sky. I have an uncommon affection for this remarkable little automobile, an ivory, 1300cc hatchback, the most basic Honda Civic, my partner for 242,915 miles when this trip began. I bought her new in June 1982 for less than $6,000. Now she was on her second clutch, second water pump, and third timing belt. The engine had had one rings-and-valves job and could have used another. The same mechanics had been working on her all her life. Any long road trip over uncertain terrain involves risks, but I felt as good about setting out in Powdermilk as in any fancy new SUV. And with the seat reclined, I could sleep in her comfortably, with plenty of leg room, when no outdoor campsite presented itself. Her ground clearance was a little low (compared to my earlier five Volkswagens), but she had taken me over a great array of abominable roads, and back. To do what needed to be done, she required little gas and a quart of oil now and then. On the way back to the highway, a rough bunch in a 4×4 looked as if they'd just seen Bigfoot when they beheld Powdermilk coming down that grade.

The road made a sudden burst out of the Blue Mountains into sage and juniper, and the temperature went up ten degrees; then solid rabbitbrush and hotter yet. I stopped for a Fudgsicle at the old store in Ironside and asked the jumbo proprietor which of two possible routes I should take. He took a look at Powdermilk and told me emphatically *not* to take the one I was inclined toward, unless I wanted to get stuck real good. Though I knew he was underestimating my rig, I took his advice.

Outside of Ironside, a fifty- or sixty-pound badger lay dead on the road. Often I stop to examine road-killed animals, and if they are fresh, I move them off the pavement onto a bed of grass. I hate to see a beautiful creature smeared over the pavement. The big badger was dusty when I picked it up, like a heavy old rug. I ran my hand over its broad, soft, striped forehead, thick dense pelt, powerful paws like backhoes built of sinew and bone. I plopped it in a cloud of its own dust among autumn straw and old irises. There had been a porcupine in the road earlier on, but I had let it lie. Sometimes a fetid carcass will be covered with butterflies, males only, come to probe the maggoty flesh for organic nutrients that the less active female butterflies seem to do fine without.

At Brogan Summit the Snake River Plain came into distant view. Willow Creek, a red-painted rocky canyon of poison ivy and broken rock, made a long descent through folded-dough hills, sorghum-red in the sunset. And so I came to Farewell Bend, where wagon trains on the Oregon Trail commonly left the Snake River behind to tackle the rugged ruts ahead. It is where I would rejoin the Snake.

Some summers past, Thea and I saw monarchs preparing for an overnight aggregation at Farewell Bend. Now I crossed into Mountain Time, and it was too late to look. I camped — really *camped* for a change — on the ground in a grove of locusts in the state park, beneath an unobscured and barely waning moon. I had now been a week without monarchs. Tomorrow I would follow the Snake into Hell's Canyon. I felt secure in the belief that my luck must soon change.

SEVEN

Hell's Canyon

When the early morning sun struck the roost trees at Farewell Bend, I walked along the drawn-down Snake's muddy shore. Nothing showed on the locusts or on the nectar below. I showered in the park, then gnawed a carrot for breakfast as a tawny satyr anglewing floated about the camp on mandolin music. The couple with the mandolin, at the next site, offered me coffee, but I was eager to be away. Since the time zone ran right past the park, it seemed as if I had an extra hour for exploration.

Huntington is a railroad town full of disused cabooses. On the end wall of an old block of brick buildings, a big painted ad, faded now, read: "Clark's Cafe / ALL WHITE HELP / A good place to eat." At the nearby truck stop the night before, my black server told me about being evicted from her white fiancé's apartment. Some things change slowly. There was still a café in this block and a bar. Farther along, a pair of cowboy boots stood in the locust leaves drifting in the gutter. They looked pretty well worn. I went in for coffee and a doughnut to back up the carrot. At the counter a skinny, pocked, and mustached railroad cowboy looked askance at my colorful T-shirt from the Tucson "Invertebrates in Captivity" conference. "Howdy," I said. "Howdy?" he replied.

I asked the waitress about the boots. "Oh, are they still out there?" she asked. "I bet those're Pete's."

In order to investigate the Snake in Hell's Canyon, I was obliged to approach from the south. The only way to follow the river out of the north would have been by water, either in a jet-boat (no way to watch butterflies) or a whitewater raft, and time and logistics did not allow for that. The Snake River roadside was painted with the vivid purple of asters and yellow of rabbitbrush and sunflower, flickering with the familiar white of cabbage butterflies. Boise Public Radio offered ZZ Top, Stevie Ray, and Janis, all of whom tended to make me drive too fast over the rippled surface. I slowed to jot a couple of captions for the trip so far: "Washboard roads of the West," and "Of cabbages & kings, mostly cabbages." Golden-mantled ground squirrels, fattening for fall, climbed sunflower stalks for their seeds. Milkweed on the bank was chewed, by no one in evidence. The drawdown of Brownlee Reservoir left bright turquoise bands along the shore like chlorine rings in an old swimming pool.

An official-looking green-and-white sign read "Welcome to Jack Gordon Oregon/Unincorporated/Population 2 or more/elevation 2096′." Jack Gordon was shady and inviting, unlike some of the barrens of fishing-boat trailers strung out in wide spots of dust and sun, where languorous, hot, windblown wives lay about in the open backs of fishermen's rigs, feet out. Fish humped up in the tepid shallows, boats glinted out on the reservoir. Hunters were just as numerous. Two guys rumbled by with a big buck trussed up in a truck. My game would have struck them as laughable.

Above a patch of blue willow and below a blue sage hill, I came suddenly into a huge flight of Nevada buck moths. They fluttered slowly, and I nearly caught two with my left hand before Marsha did the job. I was more familiar with the related elegant day moth, an orange-and-orchid wonder that shoots through your field of vision like a bright bat. *Hemileuca nevadensis* has soot-black and linen-white wings, each with an eye-spot, and a silky white thorax; its six furry legs and rear end are banded with bright orange. When I held one in the forceps for a close look, it twisted that bright butt around and around, emphasizing its wasplike, toxic-looking qualities. Dozens were on the wing at once; clearly, there must have been hundreds in the colony. When I described this dramatic irruption at the annual Northwest lepidopterists' gathering in Corvallis later that fall, I learned that there were just a couple of prior records for buck moths in the entire Pacific

Northwest. My moth collector friends freaked out when I confessed that I had taken no specimens. Not in a collecting mood, I let the handsome moths flutter past.

At the mouth of a small canyon tributary, an old orchard bore a weathered sign: "Denzel Ferguson Desert Riparian Research Station / Fox Creek Experimental Farm / & More." It appeared that more was less. The place looked kaput, a double-wide trailer open to the weather and the land gone to mega-thistles and desiccated burdock. Denzel and Nancy Ferguson are well-known Northwest environmentalists who wrote a hardball book about cattle on the land, *Sacred Cows at the Public Trough*. What dreams of theirs had come and gone at weedy little Fox Creek?

In other side draws, wooden boxcars, trailers, and temporary hunters' camps hunched against the hot rock. I stopped by a burned-out trailer in the shade of two little black walnut trees for cold water and an apple. The meats of the small nuts were very flavorful, but far too hard to extract for the time I had and the miles I meant to cover. A mourning cloak drifted among the falling yellow leaves.

I thought I heard a low jet, but it was a bird ripping past my ear. It dove into the sandbar willows and I never saw what it was. There were bright yellow sulphurs all over the purple asters, big coronis fritillaries and bay-brown zephyr anglewings on the rabbitbrush. A Milbert's tortoiseshell flew downriver, a mourning cloak up. And then — *Yes! A monarch!* — rose from the rabbitbrush brink and flapped out over the lake. Fighting a downstream breeze, it vanished to the south. This happened at noon, almost opposite the Mountain Man Lodge & Marina at Dennet Creek. *What* a relief — they hadn't all gone south already. I felt released from a week's hard penance.

Climbing up and out of the canyons, the road ran across slablike walls fractured into purple scree. The slopes looked cat-scratched with maroon stripes of sumac and poison ivy's scarlet, like venous and arterial blood. A prairie falcon skinned the ridge. The Wallowas reared up above, and the Snake ran turgid *way* down below, as the road left the canyon and veered toward the west. The Daly Creek Pastoral Company had posted the next 7.1 miles of road, meandering up the side of a grassy valley. Sunflower heads poked through a cattle guard. Pastoral green hillsides were inscribed with the hoofways of generations of grazing cattle. These parallel ruts, known to ecologists as terracettes, mirrored the drawndown lake levels of the stilled Snake: mighty strange these new curves and hems of the land would appear to the likes of Lewis and Clark.

In spite of heavy grazing, a creeklet in the crease below the road was daubed with low yellow sneezeweed, watercress, and big blue forget-me-nots. This floral host was crowded with ringlets, sulphurs, whites, coppers, skippers, a zephyr, five mourning cloaks, and even two California tortoiseshells. This orange-and-black butterfly erupts in vast numbers in certain years, engaging in mass movements around the Cascadean region before it defoliates its buckbrush host plants, its parasites increase, and its numbers crash. In its big years, these tortoiseshells swarm through the high country of the Cascades by the millions, closing roads and leading to spurious reports of monarch flights. Many people, hearing that I was following monarchs, told me about the time they had seen scads of monarchs on Mt. Hood or one of the other Cascades volcanoes. This was a crash year, so it was unusual to see any California tortoiseshells at all. These two flew up and away to the south.

I paused near the mouth of the Powder River, which used to end in a final plunge to the Snake River; now its last meander drowns in a slackwater bay of Brownlee Reservoir. The name Powder River brought back painful memories. In July 1970, my brother Bud lost two close friends and nearly his own life in a horrid head-on collision between an old Chevy and a cattle truck, upstream near the town of North Powder. For weeks I trailed butterflies with a camera and a long lens in the Blue Mountains and the Grande Ronde Valley, waiting for Bud to come to in a La Grande hospital. I remembered watching monarchs and Leto fritillaries together on thistles at Morgan Lake, to which I frequently retreated from the bedside vigil. Walking the streets of La Grande at night beneath big birches, wondering what Bud's life would be like if he lived. Hearing his first spoken word, "Thoreau," when I read him a favorite quotation: "To think you could waste time without injuring eternity." Bud survived, with his intelligence and wit intact, though his life since has been completely defined by that bad night at North Powder.

I rifled a barren patch of milkweed, then veered back north. A golden eagle sailed alongside, then nearly into the windshield, hang-gliding the ridge above the road. I drove through the isolated town of Halfway, at the base of the snow-spiked peaks of the Eagle Cap Wilderness, and climbed out of a valley spattered caramel and candy-apple red by fall hawthorns. I regained the Snake at Oxbow and entered Hell's Canyon — the deepest gorge in North America. Umber palisades began to climb skyward and close in as the narrow road entered its maw.

A patch of goldenrod by a backwater culvert was completely thronged with dozens of ochre ringlets, mylitta crescents, and Juba skippers at four o'clock. I was shocked by the late hour, especially on Mountain Time. Evening shadows come early under the steep walls of the canyon. But the sun was still staring directly at Big Bar, an onshore bottomland that seemed afloat in a yellow raft of sunflowers. The road ran above Big Bar, and I turned off to check it out. Right away I saw a monarch.

I caught the small dark female and tagged her as a second one appeared and dallied south. Then a third, a worn individual, fluttered down toward the river and disappeared. Number four was another little female, who after tagging resumed nectaring after a few yards' flight. I watched and followed when she settled onto another sunflower. Three more times she lofted, landed, and drank, as I followed her south and finally lost her over a hill. I cruised back over the flat, conning the sunflowers. Big female alfalfa butterflies, intense Union Pacific orange with yellow-spotted black borders, stacked the sunflowers, setting off a succession of false alarms.

The fifth monarch was a big one, likely a male, that I bricked as it nectared on a fat sunflower. I felt drunk on monarchs, and I grew a little silly with my netting. Like the others, he cruised off south faster than I could follow. Shortly before sunset, a sixth sailed into a hackberry as if to roost, but I could not find it in the tree. There were many others, but these were the ones I saw clearly before the shadows fell, the sun dropped over the backside of Hell's Canyon, the temperature dropped, and the action subsided.

John Eckels and Arthur Ritchie, who farmed Big Bar in the 1890s, won prizes for their produce at the Trans-Mississippi Fair at Omaha in 1898. Big Bar is now mostly inundated, but their sunflowers still won first prize from me, the monarchs, and the goldfinches that plucked their ripening seeds. Big Bar was also an informal recreation site, peopled by fisherfolk in campers. One family's black lab attended noisily to me the whole time, and when I left, its owner gave me an apologetic smile and a big wave. I wished them as much luck with whatever they were fishing for as I'd had with my prey.

Several miles downstream in sunset shade, I got another smile and wave from a black dad whose family was fishing on top of the parapet of Hell's Canyon Dam. Across and below the imposing, appalling dam stood the handsome wood-and-stone Hell's Canyon Creek visitor's center. How I wanted to give the river a mighty Heimlich, see it spit out this salmon-stopping hunk of concrete! But here below the dam now raged the *real* river, grand to see at last. And the phenomenally broken canyon, much higher and narrower than the black-basalt and red-herring rivers I'd been following for the past week. I'd seen Hell's Canyon from the Buckhorn Viewpoint above the Imnaha River and farther downstream at the mouth of the Grande Ronde River. But this was the deepest I'd been in its belly. My mental immersion in the cold, wild-flowing river was disturbed by a huge green grasshopper that appeared on the stone wall above the drop. The only other folks at the visitor's center noticed it too, and instead of discussing the large subjects of dam, river, and migration, we talked about the grasshopper.

As I stood watching the thunderous spume from the dam, a party put in for a twilight float downstream on the wild river, and a jet-boat arrived with a sputtering grumble. The proper uses of the river are a source of never-ending rancor in the Hell's Canyon National Recreation Area. Many feel that the quietude of the wild river should be appreciated by floating it, without the intrusion of roaring engines;

but jet-boats have been on the Snake for a long time and have their own vociferous constituency. I regard them in much the same way I do jetskis, the "personal watercraft" that rend the silence of many public lakes and waterways. It was put best by Robin Cody in *Voyage of a Summer Sun,* a compelling tale of canoeing the entire Columbia: "The whine of jet skis shattered the thin air. Two grown men, their wetsuits filled to capacity, throttled past and unzipped the river. I rocked in their wake. I smelled their fuel and heard them round the bend. Waiting for birdcall to come back to the river, I reflected on the whole idea of jet skiers and why we should let them live."

It was back in 1975 when I first saw the power of the Snake River to deliver wandering butterflies into the Northwest. Five years after his accident, my brother Bud was able to accompany me in the field on a summer survey of butterflies in little-collected parts of Washington. On the last day of July we visited the confluence of the Grande Ronde and the Snake at the very mouth of Hell's Canyon. Viceroys flew, and we even turned up a natural hybrid between a viceroy and the related Lorquin's admiral. But the great prize was a series of seven males of a diminutive species known as the dainty sulphur or dwarf yellow: the first *Nathalis iole irene* ever recorded in the Pacific Northwest.

When people regard the monarch and its amazing long-distance flights, they often imagine that only a big, robust butterfly could carry out such a feat. Yet the dainty sulphur, whose wings wouldn't cover two average thumbnails, also flies far. A resident of warm, southerly climes, *N. iole* pours into the North each spring, in most years filling the center of the continent with breeding butterflies by the Fourth of July. Like monarchs in spring, they leapfrog north in successive, quick generations. They cannot survive the frosts, nor do they return south in the fall. But by invading the continent each year, dwarf yellows are in position to exploit mild winters or a warming trend by surviving farther north than usual. It is also possible that they will evolve a winter diapause, as some other sulphurs have, to get them through the cold. But this can occur only if they push the envelope of their cold tolerance each year, and push it they do.

But dainty sulphurs don't normally come to the Northwest or the

Northeast. So when we found those seven fresh males on the Grande Ronde, I was not only surprised but amazed. Later I would discover a few Idaho and Montana records, and two or three very old ones from the Mid-Atlantic states. All of these occurred in big years for *Nathalis iole* immigration, and all were found along rivers: the Clark Fork, the Susquehanna, the Hudson. So in years of great abundance, dainty sulphurs leak way out of their normal range, following the vegetation arrayed along rivers. I had apprehended the importance of water-courses for butterfly movements ever since, as a young collector, I found mountain species on the High Plains and prairie butterflies in the foothills along the High Line Canal. But this tiny flier really showed me how rivers can order the distribution of far-flung butterflies across the landscape.

The concluding arc of my long day's circle was a curvy, up-down, in-out drive east and south through Cambridge and Weiser, Idaho. As night deepened, wildlife came out. A big buck that had escaped the brigades of hunters so far almost became a hood ornament for Powdermilk, followed by three near-suicidal owls: a great horned sitting in the road like a stone, a little screech fluttering at the windshield, and a barn, ghosting alongside for a quarter mile before veering safely away. Back in the sack and sound asleep at Farewell Bend, I heard piggy snuffling and catty scratching near my head. Porcupine? Skunk? Badger? I didn't even *move*, let alone look. The interloper loped off, and I lay looking at the stars and recalling the sight of sunflowers graced with monarchs in the late sunshine. The next thing I heard was a particularly persistent killdeer committed to waking up the whole park. It is not called *Charidrias vociferus* for nothing.

I got up, wet my face, and walked the shoreline, eager not to miss any waking monarchs. Sunrise blazed off the line of locusts, but the next couple of hours revealed no departures, except more killdeers and dozens of western meadowlarks migrating along the Snake. Spotted sandpipers began to fly off with the killdeers, then sorted themselves out. Migrants sometimes mix with others of different species, as with the famous waves of warblers in the East, or the various nymphs I had seen up on Dixie Summit. But usually like migrates with like.

On the picnic table I laid out a pre-dam U.S. Geological Survey topographical map of the Snake Basin and followed the long river's loopy course with my finger. Back at its mouth, at Sacajawea Park, I had watched a monarch shoot out of the Snake and cross the Columbia — unaware that it would have to cross the river again eventually to get south — and wondered whether some Columbia Basin monarchs might not carry on southeasterly down the Powder River slot to meet the Snake just about where I had seen the first one yesterday, at a point just over one hundred miles due southeast of the Great Bifurcation. Here they would coalesce with monarchs originating all across the Palouse country, that interior upland hedged by the Spokane River to the north and the Coeur d'Alene/Salmon River country to the east, then funneled southward into Hell's Canyon.

Later findings supported these conjectures. The following fall Thea and I would visit the confluence of the Grande Ronde and the Snake, near where the dainty sulphurs had turned up. Between Limekiln Rapids and Captain John Rapids, we found lots of monarchs nectaring on the cerise spikes of purple loosestrife blooming in a riverside seep. Once tagged, they charged off south into the mouth of Hell's Canyon. The next day's travel would likely have taken them most of the way to the other end of the gorge, where they might well have fetched up among the sunflowers of Big Bar. Soon thereafter, David Branch and I tagged bunches of monarchs outside of Pendleton, Oregon. They were moving up gullies that pointed over the Wallowa/Blue Mountain divide toward the Grande Ronde, the Powder, and ultimately the Snake. I didn't know any of this as I pondered the map at Farewell Bend. But if I had, it would only have reinforced my feeling that the Snake River corridor is a powerful migratory magnet for monarchs as well as meadowlarks.

Mandolin music came again from the next-door campsite, and this time I accepted a cup of coffee and joined in on "Shenandoah" on my mouth harp before I was bound away. Packing up camp, I found a beautiful orange orb-weaver with yellow horns on her abdomen and her web strung to the ground from Powdermilk's bumper. I transferred her to the nearby brush, as it was time to go. I had my running orders once more.

EIGHT

Snake

In Mexico, the Day of the Dead celebration on the first of November is celebrated with sugar skulls and marigolds — lots and lots of marigolds. Another symbol of the spirit of the departed, especially in the state of Michoacán, is *la monarca.* A second name for monarchs in the villages near the winter clusters is *las palomas,* which means the doves, or the souls of lost children. When the monarchs come back in late fall, they are seen as the souls of those who died young, returning for their annual visit. I remember one Day of the Dead altar in a tomblike space beneath the streets of Mexico City, near the Diego Rivera museum. Thousands of real marigolds mingled with thousands more made of orange tissue paper and just as many monarchs made of crepe, all wrapped around a city of skulls. Now, as I stood among a hundred acres of marigolds beside the Snake River, masses of orange and a piercing pungency on the air took me back to that spectacle.

This morning, coming from Farewell Bend, I followed the Snake to Weiser, then drove south on the Oregon shore. The river flowed broad and slow and smooth across a flat world of agriculture and weeds. From burning fields wafted an odor that evokes autumn anywhere. A sandy prairie falcon shot low over fields of yellow onions ready for harvest. A blond woman and a boy on a big white horse walked the milkweed ditch. Deer Flat National Wildlife Refuge, to judge from the map, was a thin riparian strip in the middle of the onion fields, but I couldn't get to it. I plucked an onion from the road for a future meal, and at a produce stand I bought a buck's worth of Jonathans and little tomatoes to go with the onion and the last of my cheese and bread.

I ate them at the mouth of the Malheur River. Wapato, an arrowhead-leafed emergent plant with delicate white flowers and starchy tubers, much valued by the Indians, still grew in the backwaters. In 1864 W. H. Packwood ran a ferry here with six oarsmen, taking in up to a thousand dollars per day in fees. Now nearby Nyssa has a bridge and signifies itself "The Gateway City, Where the Oregon Trail Enters Oregon." Annual revels include Thunderegg Days, the Nyssa Nite Rodeo, and the Catfish & Crappie Carnival. But row crops are the big thing, with 800 jobs locally during the sugarbeet harvest, a million cases of corn canned each year, and 20 percent of the nation's onions grown in the outlying "Treasure Valley." This may sound wholesome, but the downside is the heavy pesticide load carried by local waters, fields, and folks. In Nyssa I scanned a large bed of zinnias, a monarch favorite, but sprayed fields spread for miles around and *no* butterflies showed, not even cabbages.

Some miles on, I worked a railway embankment of rabbitbrush and aster above the tangled floodplain of the Snake. Then, gazing across the hazed bottoms, I made out yellow and orange smears that had to be fields of sunflowers and marigolds grown for seed. I found a road to the fields and tried to ask permission at the nearest house, where I found lots of cats but no people. Scratched behind the ears, the cats said fine, go right ahead. Anyway, I reasoned that flower farmers must be gentle, tractable folk, environmentally sensitive and friendly to butterflies and those who would study them.

So I drove down a tractor path to a cornfield and ditch, then continued on foot to the immense fields of flowers, a great colorful smorgasbord for butterflies. I took a long walk on a farm road all around the perimeter of this vast quilt of petals — many acres each, not of sunflowers but of gloriosa daisies, other yellow composites, bachelor's buttons, multicolored big asters, hollyhocks, purple coneflowers, and, yes, marigolds. Most of the *Echinacea* were past, a pity, since I have seen fields of coneflower almost clogged with nectaring butterflies, but the other plants were in full bloom. The afternoon was hot and sticky, and masses of still another species of caddis fly — tiny and mottled — used my head as a portable swarm-marker. Mosquitoes, too. Down at the river a young couple was fishing from a boat, and I heard voices speaking Spanish upstream.

A track had been bulldozed to the river right through the riparian vegetation of the "refuge." As I headed toward the river, a truck bore down on me with three cowboy hats backlit in a corona of dust. In the truck were two men and a woman, maybe twenty years old, wearing identical ranch-logo T-shirts. "What the hell are you doing out here?" the driver demanded, more bemused than hostile. I said I'd tried to ask permission and explained, "I'm likely to find some monarchs here."

"I doubt it," he said.

"I'll just go past the marigolds and head back to my car," I said. "Okay?"

"Well, keep walking then," said the lad. "And I hope my grandpa doesn't see you. If he does, he'll shoot. He'll have my butt, too, and he'll shoot *you* right down."

They barreled off in a dust cloud, and soon I heard their shouts and whinnies down at the riverside. The only words I made out clearly were "Fuckin' butterflies!"

And that's how I came to be standing in a world of marigolds and thinking about the Day of the Dead. From the shade of two enormous green John Deeres, I scanned at length across the expanse of orange, looking for even oranger flickers of movement while keeping half an eye out for Grandpa. I never saw him or monarchs, either, and on my way out I noted that the land wasn't even posted or fenced. By sunset I reached the mouth of the Owyhee River, then crossed the Snake and left Oregon behind for Idaho.

I finally found some of the Deer Flat Refuge beside Lowell Lake, south of Caldwell, where I camped. In the morning I learned that the refuge was closed for the season. The 7,200-foot dam was topped by a handsome stone parapet of purply basalt blocks, built by the Civilian Conservation Corps in 1938 and rebuilt by the Marsing Junior CCC in 1975. A carful of Marsing's present-day teens pulled in, playing loud music and drinking sodas. One girl popped into the brush for a pee while another mugged on the parapet and the driver honked. They waved my way and honked again as they left, with me now thoroughly awake. Slipping past the fence, I took a walk along the beach in the frosty morning, listening to the two-note calls of elegant, tuxedo-toned western grebes offshore.

A radio program aimed at long-haul truckers said it was supposed to be 84 here today, and that Highway 287 was closed because of a whiteout in Colorado. At home there were high winds and floods. I passed a great field of asters — purples, pinks, reds, and whites. It was devoid of butterflies, perhaps because of pesticides or the night's frost — but I still felt I was in the right place.

Not even a sulphur showed, though alfalfa fields were nearby. Alfalfa usually abounds with sulphurs, which are far more common today than ever before because of the widespread cultivation of the leguminous hay crop the Europeans call lucerne. Both the common yellow sulphur *(Colias philodice)* and the orange sulphur, or alfalfa butterfly *(C. eurytheme)*, feed happily as larvae on alfalfa, white and red clovers, and other exotic legumes. This has caused such unnatural abundance of both species, side by side, that their barriers to hybridization have broken down, and many individuals now show mixed traits. Orley Taylor, later director of Monarch Watch, discovered this. Lepidopterists speculate that the two species might one day merge, though for now there is still a prevalence of pure examples of each.

Another insect looms large in alfalfa culture wherever native pollinators have been reduced by agriculture. I was seeing big blue boxes on legs and red legless ones the size of freezers set along many fields. They baffled me until I stopped into V & F Produce, bought some Ida Red apples and frying spuds, and queried a saleslady. "Oh, those are the houses for the leafcutter bees," she said. "They like it when the days get up to a hundred degrees. Different folks have different theories as to what works best, whether the bees prefer red or blue houses, north or south facing, on legs or not, and so on."

Leafcutting bees nest singly in holes or tubes. Each bee box contains thousands of pencil-sized breeding tubes. The bees cut disks of leaf material, line and cap the cavities with it, and lay their eggs, one per cell. They lay down a pollen/honey paste to further provision their young. In visiting plants to take leaf or petal bits, pollen, and nectar, they effect pollination, and the slices of leaves they take are more than made up for by this service. In 1930 the Eurasian species *Megachile rotundata* was introduced especially to aid in pollinating the alfalfa crop. It takes around ten thousand bees to pollinate two hectares of alfalfa, hence the big boxes. Shipping and providing these bees has become a substantial business in alfalfa territory. Their presence has to be carefully timed with the bloom and coordinated with spraying schedules in the area. Of course, native pollinators of native plants have largely been eliminated locally, as the absence of butterflies in the flower fields likely demonstrated.

I asked the fruit-stand woman if she ever saw any monarchs. "We get a few, but never a whole lot," she said.

I rejoined the Snake at Marsing, which seemed a fairly typical past-peak farm town, with video rentals and mini-marts instead of an old movie house and a general store. But it also had a "Tanning & Book Store" and "Guns & Ammo, Books." A colorful carved sign at the edge of town said "Marsing Idaho Welcomes You to the Valley of Fruit and Harvest, Gateway to the Owyhees." An arched mural depicted pioneer oxen and wagons along the fruited river. The shore below the sign was a gravel boat launch, where the head and eviscerated front half of a big pug-nosed fish, its eyes dusty marbles, swam grisly in the dust. A pretty litter of red and white onion skins blew everywhere, like bits of silky Japanese paper.

"Normal snow removal 6 AM–6 PM," read signs along the highway. "Travel at your own risk Marsing to Bruneau." I took that risk. Speeding up into the country, past more of the red and blue bee boxes, I thought about the lengths we go to in replacing lost pollinators. That brought to mind Douglas Whynott's riveting book *Following the Bloom,* which describes migratory beekeepers who haul semitrailer-truck loads of honeybee hives from Florida citrus groves to Maine blueberry barrens to California almonds to North Dakota sunflowers. Accepting bee stings and hard knocks while evading truck scales and quarantines, they lease hives and sell honey to make an unpredictable living in constant pursuit of nectar and pollen.

While the beekeepers' migration is both commercially driven and artificial, built around alien insects and domestic crops, it is nonetheless dependent upon the natural phenology of the land. As both the length of the day and the nighttime temperature diminish, the plants drop leaf and bloom, and the animals must respond with them — shifting diet and growing heavier pelage, hibernating, migrating. So it is for the migratory beekeepers, and so it is, exactly, for monarchs. So it was even for me, following the sun to the south, for neither profit nor survival, but linked to the seasons just the same. And, like any migrant, I felt I had no choice in the matter.

I traveled the Snake along a narrow irrigated green strip, greener beside the river, between blond, brown-capped mesas on the Idaho side, blocky mountains off in Oregon. Hundreds of barn swallows swooped over the water, raking caddis, stoking for their big trip. The riverside vegetation was now largely tamarisk and saltbush. At Bernard Landing I investigated the saltbush and, as I hoped, found pygmy blues. In a cove of *Atriplex,* scads of them flickered about, the smallest butterflies I had ever handled, so tiny that a monarch tag would completely cover one of them. They were bright blue at the base of the wings, rich bronze beyond, with brilliant silver scintillae bordering black spots below. Now I knew I was in the biologic, if not yet the hydrologic, Great Basin, with saltbush and *Brephidium exile.*

Plants tolerant of the alkaline soils of the Great Basin are called halophytes. Saltbush is one; another is salt cedar, an introduced shrub also known as tamarisk. I tried to push through a forest of wispy tamarisk brush, but its lower branches grew tangled and I was repelled. Salt cedar has captured thousands of miles of southwestern shorelines, sucking up an awful lot of water while usurping the riparian zone from willows and the birds dependent on them. While I deplore its ecological impact, I love its cotton-candy florets with their soft sweet smell, and I owe it a memorable afternoon in its native range.

In 1978, in Turkmenia, I was traveling by boat on the Kara Kum Canal, which runs between the ancient city of Mirv and the Caspian Sea. Tall reeds walled the banks, blue-cheeked bee-eaters crossed between them, and camels came down to drink. The warm Russian beer was bad, but in the bow, the desert air was fresh. After a shady lunch ashore on Ashkabad carpets with thick, pruny wine, I wandered into a native stand of pink-clouded, ten-foot tamarisk. All of a sudden, in the pale blue above me, appeared the first Asian monarch I had ever seen. *Danaus chrysippus,* its wingtips both blacker and whiter than a monarch's, floated on the air for a long look before sailing off down the canal. Now the Snake was just as sluggish, the Idaho air as dry as on that afternoon in Asia Minor. But instead of an American monarch against the sky, I saw only cabbage whites at the tamarisk blooms.

At Melba Crossing an Idaho historical sign told the story of the steamer *Shoshone.* Built in 1866, it plied the Snake between here and Olds Ferry on the way to Boise. Once it explored sixty miles upstream, where steam had never gone, "astonishing the jackrabbits with its ambitious whistle." Business was poor and firewood for the boilers scarce, and after a few trips, service stopped. "Finally, in a hair-raising ride, the 136-foot, 300-ton boat was run down through Hell's Canyon to the Columbia."

Across the bridge an Idaho-shaped stone sign placed by the Daughters of Idaho Pioneers told of James Cox, ambushed and killed by Indians on October 25, 1865, on the Boise–San Francisco stage. His wife escaped and drove their wagon to the ferry. Bigfoot, leader of the "outlaw Indians," was killed by J. W. Wheeler near here in Reynolds Creek Canyon, while waiting to rob the stage, in July 1868. And on July 31, 1878, William S. Hemingway was shot by Indians while driving the stage to Silver City, but he got the stage to the ferry here, "where he expired."

Things had slowed down in Idaho. Murphy was nothing more than the Owyhee County Museum, the courthouse, a few trailers, and a closed shop. Beyond, the vast Snake River Plain rose into the heat-rippled distance. To the north lay the Snake River Birds of Prey National Conservation Area, with the densest breeding population of raptors anywhere. I saw northern harriers, kestrels, prairie falcons, redtails, roughlegs, and raptorlike ravens. Butterflies, too, were abundant. Big lonely rabbitbrushes were plastered with ten species, especially pygmy blues and great white skippers. Beside a fetid ditch, hundreds of brilliant orange-bordered blues glittered over an alfalfa field. Moths mobbed the flowers, too — tigers, millers, grass moths. A spotted lizard waited below for one to flop near, heavy with nectar. When I tried to catch the lizard for a good look, it left its tail in my fingers. I'd forgotten about that snappy little adaptation, and felt bad as the tail whipped back and forth on my palm. I hoped it would grow back before it was needed for real.

After many miles without a stopper, the river now emerged from C. J. Strike Dam. A beautiful clearwater canal, lined with goldenrod, sunflower, rabbitbrush, and aster, rushed out of the reservoir too, but up at lake level, paralleling the Snake, which issued from the bottom of the blockage. As I was looking over the milkweed in the opposite field, I heard a funny chuckle-snort behind me. Swiveling, I saw a river otter's head pop up in the water, then another, another, another. Four sleek and serpentine otters were looping upstream in the canal, rising, diving, munching something. Their fur shone rich brown and silver in the sun. I followed along, becoming the fifth otter in my mind. I felt the cold Teton meltwater parting over my pelt, my guard hairs glistening in the sunstruck ripples, the hot afternoon and my too-human sweat falling away beneath the cool, clean flow of the surface. I rose, snorted, and climbed out from one of the most wonderful otter plunges I have ever had. The four continued up the canal without me, snorting and diving, making trails on the surface, then erupting through their own bubbles and starting all over again.

Reluctantly, I left the otters to check the park below the dam — just a lawn with shade trees smelling of insecticide and an herbicided pebble barren fronting the earthen span itself — no butterflies. Nearly reconciled to another monarchless day, I returned to the canal to look for the otters again. There was no sign of them. At five I was driving back and forth along the canal looking for otters, when a bright butterfly appeared on the bank. I leapt out with Marsha in hand and watched the monarch glide south across the canal to the goldenrods. I followed a little way along the bank, but lost it in the sun just as I was getting my binoculars on it. Neither the monarch nor the otters showed again. But along the milkweed-lined road from the dam-side village back to the highway, another monarch crossed my path from the south. Circling first, it flew into a bushy peach-leaf willow, where I presumed it would roost overnight, but I couldn't spot it. I would have camped there, but for one of the rare social commitments I allowed myself on the trip. I left the monarch in peace.

As I crossed the Snake at Bruneau, mayflies glittered and swirled in the falling sun. "Would the last person to leave the Bruneau Valley please feed and water the protected snail?" read a billboard on the edge of town. I passed Wendell, where a newspaper photographer and monarch watcher named Steve Kohler saw an overnight bivouac of sixty migrating monarchs clinging to a cottonwood branch in September 1991. Would I see such a thing?

I arrived late in Filer for my brother-in-law Leon Martin's birthday. He and Thea's sister Anne both work long hours in health care, and Leon was already sleepy. Back in Ontario, Oregon, I had bought two locally made knives, good steel with simple wooden handles, one for cutting up my onion and potatoes, the other for Leon's birthday present. He liked it, but reminded me that giving a knife is bad luck, so he bought it from me for a penny. Anne cooked a steak for our brief visit, and we turned in. The night was hot and close. Before I dropped off, I recalled our last visit with Anne and Leon. He had loaned me a helmet and taken me down across the Snake's broad canyon above Twin Falls on his big Suzuki motorcycle. We crossed below gushing Shoshone Falls, near where Evel Knievel had tried to leap the gorge on a motorcycle and crashed. Fragrant rain in my nostrils, the air rushing by in the green, rocky canyon, I had imagined myself chasing monarchs there. And now here I was, minus the motorcycle.

On the first of October, the shade trees of Filer and Buhl were all gold. A big wooden monarch hung on the side of a stone-built house, and I wondered if I could count it. I like a district where you can read the geology in the dwellings. Many buildings here were made of local basalt quarried from the lava flows I had been following ever since the coulee country in Washington. Now I came to one of the most dramatic features of the entire western lavalands: multiple waterfalls bursting directly out of the basalt walls and plunging to the river — the Thousand Springs. The Lost River, surface water that disappears below the lava flows to the northeast near Craters of the Moon, is thought to reappear here some 150 miles and 150 years later.

Of course, the springs are not what they were. Everything has changed since the whites came. In 1812, at nearby Salmon Falls, Joseph Miller found one hundred lodges of Indians spearing thousands of salmon each afternoon, to dry for winter. In 1843, John C. Frémont described the spearfishers at Fishing Falls, the Snake's highest salmon cascades, as "unusually gay . . . fond of laughter; and in their apparent good nature and merry character . . . entirely different from the Indians we were accustomed to see." He marveled at Salmon Falls' eighteen-foot vertical drop, adjacent to "a sheet of foaming water . . . divided and broken into cataracts" by islands that "give it much picturesque beauty, and make it one of those places that the traveler turns again and again to fix in his memory." It is still pretty nice to look at, but the last salmon spawned here in 1910. And while the view of the Thousand Springs is still one to be fixed in memory, their flow has been much diminished by groundwater extraction for irrigation and hydropower, so now it is more like the Dozen Springs.

On my way from Hagerman to The Nature Conservancy's Thousand Springs Preserve, crossing Billingsley Creek, a monarch shot ESE across my bow. I saw it well before it vanished in the general direction of the Snake's upstream meander. Powdermilk negotiated a worm bed of little roads that led to a side canyon going down to the preserve. An old hydroelectric plant was stuffed into the draw. A line of small springs issued right out of the rocks into a super-clear channel of bright green trailing plants, lined by goldenrod and pale aster; the water then flowed under the road and into pumps and pipes with clicks and gurgles. Steel and wooden pipes ran all along the cliff to old machinery. A few larger cataracts still free-fell through the foliage of red creepers to the Snake.

I parked and walked over a pretty basalt bridge to Ritter Island, the local TNC headquarters. Minnie Miller, a Salt Lake City financier, had set up a world-class stock farm here for Guernseys, and a stone monument wreathed with creeper commemorated "Idalia of Edgemoor B. April 27, 1918, D. Dec. 1, 1934. HERD MOTHER." Minnie's beautiful two-story stone house occupied the shadiest spot on the seventy-acre island.

Chris and Mike O'Brien, the preserve managers, kindly invited me to stay in the stone house any time I wanted to survey the preserve's butterflies, asking for only a species list of what I found. I began on the Columbine Trail, on the mainland, which passed beneath a big waterfall. The day was *hot,* and I was tempted to duck under the spray, but restrained myself. Canyon wrens, crickets, and quail made the day sing. All of a sudden, sagebrush was in bloom — a sure sign of October. A golden eagle sailed over the next waterfall as a redtail crossed its path; they soared off together. The waterfall dropped into watercress and rainbow, forget-me-not and asters, boulders and crystal streams. Moss and algae, almost unheard of in southern Idaho, shimmered Oz-like. The cool mist was almost as much of a godsend as monarchs would have been. A bright little west coast lady lit on a screaming yellow ash tree, and when a white came along, looking blue in the shadow and the sky's reflection; the lady chased it into the sun, white again.

In the last of the heat I walked over Ritter Island, the furrows for native grass restoration making tough footwork. At the north end, in a line of locusts, massed brown pods of locust seeds ignited by the slanting sun resembled monarch clusters. I disturbed four sleek black-crowned night herons and a red-shafted flicker that flashed monarch-orange. Hawthorn berries, foliage, and bark all reddened the bank. A *Danaus* here, master of self-advertisement, would prove almost cryptic against the autumn pigments.

A side path labeled "Minnie's Falls" led me to a Russian olive glade by a backwater pond with a clear view — one of the most refreshing spots I have ever been in. The springs most beloved by Idalia's mistress spread across the canyon's shoulders like white tresses braided with ribbons of green. Squawbush, a harmless member of the poison oak and ivy genus, daubed the Snake River banks with nosebleed streaks. The crenellated, white brick 1913 power plant rose castlelike in the spray across from Minnie's blue barn. Two magpies flashed from castle to barn. A country tradition says a pair of magpies means good luck, and it was, just to be there.

The following June, I took Thea to see Minnie's Spring. As a storm closed in, a female monarch blew by, then settled on a floppy milkweed leaf and deposited a minute green egg on its underside. That reinforced what I was thinking now, that Thousand Springs should be a superb stopoff point for any monarchs working their way up or down the Snake. Before I left, I urged Chris to encourage milkweed and nectar plants, as I was sure monarchs would breed here in good numbers as well as stop in to stoke up.

Walking back toward the stone house, I heard an electric guitar playing chords from an old Doors song, and found the preserve foreman's son, Jeff, sitting on his back porch with a rounded redhead named Todd. They were wide-eyed at an old guy with white beard, long hair, and butterfly net, who recognized their music. Jeff's dad, Ken Pressley, came out and gave me a glass of hundred-and-fifty-year-old Lost River water. As I drank it, a rough-legged hawk wheeled over, deep into its own migration.

For now, it looked as if my leaders might continue coming in ones and twos rather than swarms. But the monarch I had seen on the slopes above, perhaps fresh from the spring-fed asters, was all I needed to set the compass needle for the next morning. For my purposes, one good pointer was worth a fieldful of wanderers who couldn't make up their minds which way to go.

Idaho, it turns out, has furnished important data for western monarch migration studies, yet the ultimate destinations of its migrants are still poorly understood. In the early days of monarch tagging experiments, Fred Urquhart caught and tagged butterflies in various places where they were common, shipped them to places where they were not, and instructed collaborators to release them. A few Idaho and British Columbia releases were recovered in California. Though common sense tells us that such transfers cannot illustrate what native migrants would do, Urquhart's maps helped promote a popular belief that all monarchs west of the Continental Divide head for the coast at migration time. Since then, a few people have persisted in conducting such outdated transfer-and-release exercises. Because of this history of manipulation, I wanted to know whether any western tagging records at all were based on wild, nontransferred monarchs.

It so happens that two Idaho women, both schoolteachers, had indeed tagged local monarchs in the Boise area and experienced recoveries in California. Faye Sutherland was known and honored all over Idaho as the Monarch Lady, for the thousands of students she inspired over three decades by bringing monarch larvae into the classroom for rearing and release. A tall, sturdy woman with black hair and a firm demeanor, she led a successful campaign to have the monarch officially declared Idaho's state butterfly. From the mid-seventies to the late eighties, at least twenty of her indigenous monarchs were recovered up and down the California coast — the first nontransfers ever recorded from outside California.

Across Boise, Mary Henshall, the daughter of Japanese-American farmers who is now a demure and elegant eighty, also taught with monarchs in her classroom. One of her techniques was to let a number of larvae pupate on a branched twig, creating a miniature monarch tree containing dozens of fresh chrysalides. One monarch she tagged was found in Ventura, another in mid-California.

So these teachers provided the first documented records of non-Californian western monarchs moving to the coast — a pattern that had been assumed for decades but without proof. While this certainly did not prove that all monarchs originating in or passing through Idaho migrate southwest, it demonstrated that some do. However, in the same year as her Ventura recovery, 1989, one of Mary's monarchs was found in Orem, Utah, far to the east, south of Salt Lake. And one of Faye's Boise releases turned up near St. George, Utah, also way south and east of Idaho. These recoveries came in a region where almost no one is *looking* for tagged monarchs, in contrast to California, where hundreds of people do so every fall and winter. The flights of these monarchs conformed more with the SSE vanishing bearings I had been observing. So while a handful of reliable, nontransfer-derived data now exist for western monarch movements, they still fail to uphold the hoary assumption that eastern and western monarchs split at the Continental Divide as if they were water itself, ineluctably drawn toward either the Pacific or the Atlantic but never both.

The Malad River was a black and blue shadow shooting across the sage steppe, streaked with yellow willows. Passing through Gooding, I eyeballed the marigolds and the tall bright asters, cousins of the wild ones I'd been seeing all along, that punctuated stoops and alleys. Beyond the city limits, irrigated fields rolled along the lower hems of the Snake River Plain. Hostile lava lumps reared off to the north. Here, green alfalfa fields were getting a third crop, fat cows rubbed flanks with sleek horses, canals brimmed with Snake Aquifer water and their banks with that potent mixture, milkweed and rabbitbrush. It should have been a fine region for monarchs, though several people had told me that Idaho populations, after a good start this year, had either collapsed or left early.

I had often ridden the Pioneer, the Amtrak train whose line ran right beside my current route, and at a crossing I stopped to gaze at a compelling scene I had missed from the train window, since it went through at night: a handsome tiered, shingled, and gambreled farmhouse, painted a cool lime green, and a wonderful tall basalt masonry barn with a square wooden tower built into the center. One bay, two buckskins, and a palomino gnoshed in the shade.

Trains loomed large in the development of this country. In Shoshone, the early optimism and pretension of the railroad town were obvious in attractive black-lava mansions and public buildings; the tracks ran down the middle of the main street, once one of the widest (and wildest) in the West. On the north side of town stood a grand old cedar-shingled building, gabled and mullioned, with a big metal and neon sign reading "Hotel Club." Also known as the McFall Hotel, it was built in 1896 and is on the National Historic Register. The next-door Columbia Hotel, its pressed-tin front radiating the midday heat, also had two names: on the side it said Hotel Shoshone. "Ernest Hemingway did some writing here when he was fishing on the Little

Wood River," said Rob Blane, who owned the McFall Hotel with his wife, Kathy. "In its heyday, the place housed movie stars and presidents." The hotel was open while the Blanes restored it, but the clientele had changed. "I take in a few nightlies if they're too drunk to get home," Rob said, "and we've got a few residentials."

Next to the McFall stood the train station, draped with reddening Boston ivy. Its fancy geometric brickwork, green half-timbering, and scrolled and capitaled pillars and pots surrounded a terra-cotta Union Pacific shield. But the building was boarded up; the new depot was a grim little fiberglass "Amshack" in faded red, white, and blue. I'd successfully avoided most news for weeks, but in a café newspaper I'd read that Congress had passed a budget containing $22 million to stay four death-sentenced Amtrak routes for six months, including the Pioneer.

"U.P. is letting it fall down," said Rob of the old depot. "You could get it for a song, and the money's there from the city and county to restore it. But U.P. wants it moved eighty feet back — it's too close to the right of way, and they don't want the liability." A long, low building, it would be a tough move. And then what would you do with it? Rob was hoping to reopen his bar and restaurant soon, and upgrade the hotel, but if the Pioneer didn't continue, it would be tough to make a go of it.

Outside town, lumps of lava were piled high along the channel of a big canal dug right out of the basalt flow, a ready quarry for Shoshone. Now the listed mansions, banks, and bars were behind me, and the only buildings were triangular, sod-covered, half-buried sugar beet warehouses poking surreptitiously from the brown surface. It was all sugar beets and sheepfold for miles. This was the time called "campaign season" in beet country, when conveyor belts pile up vast mountains of the rocklike tubers before your eyes, like a time-lapse movie of orogeny. The road was full of sugar beet trucks; fallen off, the stolid roots lay in the road like forlorn heads. An old ranch woman, in jeans and cotton shirt, her gray hair tucked under a red cap, checked each truck in as it delivered its campaign contribution to the conveyor. Just beyond the beet yard, a sign pointed doubtfully: "Ace of Clubs Bar and Blue Cow Antiques — open by chance."

On to Minidoka, another railroad burg, not so grand as Shoshone. Grand, however, was the word for the big sheet-metal water towers where the fiery locomotives not so long ago had drunk. Now their spired, shingled tops were caving, their welds rusting in parti-colored patterns. If I lived there, preferably in a basalt tower, I would paint the water towers and listen to the Mexican music flowing from the run-down little railroad shack with a new Honda outside. Gandy dancers swung sledge in the hot October sun: what must August be like? Or January? There is always a tiny park in these little towns; always blue spruces and a killdeer.

Minidoka Dam, where I next saw the Snake, was built in 1906. Teddy Roosevelt made a 20,699-acre national wildlife refuge on Lake Walcott three years later. On the undammed downstream Snake I actually saw rapids, the first since Hell's Canyon. Masses of white pelicans, cormorants, gulls, and mergansers fished in the whitewater in the steamy-hot afternoon. The stretch below was a spilled paintbox of squawbush, poison ivy, and Virginia creeper. I crossed the Snake on the longest single-lane bridge I have ever seen, at least half a mile, and met no one. The brushy islands opposite mirrored the flaking yellow wooden railings. Once across, I wandered for an hour through a maze of ambiguous little roads, fields, and crossings of the South Side Canal, trying to find my way to Burley. When I finally squeaked under I-84, hit State Highway 81, and came alongside the Snake once more, I was practically at its "bottom."

Just as, in Sacajawea Park, I had doubted that monarchs would follow a loop of the Snake northward in order to go south eventually, it seemed unlikely here that butterflies working the river would hew to it much farther. What I called its "bottom" was its southernmost extent. After following it out of the northwest for many miles, I now watched the river course turn northeasterly, toward its headwaters in the distant Tetons. If monarchs had indeed followed the Snake here from Hell's Canyon or any other points north, they would have to bail soon for the south or face winter's threat. Arcing my way across the

Snake Plain, regaining the river at Burley, I had hoped to find monarchs swept up in the net of its final swing southward.

Spotting a colorful garden as I drove toward town, I turned back and nosed into the driveway. There were marigolds, frosted butterfly bushes, asters, and other fall flowers. A patch of milkweed, beside a sign that said "Springdale Repairs," looked cultivated. At the back door, an affectionate orange kitty wanted dinner. Trying not to get shot, I peered into the backyard and saw orange butterflies playing in the sun. There were three or four of them — clearly west coast ladies. But wait . . . one of them was *big* — it was a monarch! Here at the nadir of the Snake River curve, a monarch commanded this great garden, tussling with Vanessas in the low-angle sun.

I couldn't rouse anyone, so I rushed back to the garden. It was still there. Butterflies were flying all around me, one of them the monarch, clearly joining in the pre-evening revels that painted ladies are famous for. As it circled, it looked as if its wings had been clipped. Maybe it was a nonmigratory monarch of the summer generation, a parent of the new migrants, having bred in this very garden. Or perhaps it was a new arrival, migrating itself, bird-struck along the way. I felt constrained by my obvious trespass but couldn't resist taking a swipe, and I missed. I didn't scare the monarch, but the ladies kept after it like blackbirds mobbing a crow, and it flew off toward an ash tree in the direction of the Snake. I couldn't find it roosting in the ash; it might have gone on down to the riverside brush. The ladies played on, and cabbage whites too, among rows of succulent cabbages.

The insistent cat made me suspect that someone was coming. Sure enough, a car came into the drive, with a couple coming to feed puss. The woman did a good imitation of not looking too surprised at finding me in her sister's backyard. "Here, kitty!" she called. She scratched the excited cat and poured the kibbles. "That's right, Nermal, here's your food." Then she attended to me. Yes, her sister gardened for butterflies, as I had guessed, and raised milkweed and butterfly bush especially for monarchs. "We just let them fly after they hatch," she said. "We don't know where they go. We saw some just last week, two or so." I thanked the tolerant in-laws and left.

Evening fell in Burley, a place I'd been both in blizzards and in summer drought. I took a long walk along the barbered shore of a park and across the Snake on a traffic-ridden bridge, watching for monarchs bedding down for the next day's flight south. Shrieking jetskis ripped the river. Teenage boys in the park tossed a football, struck poses with babes, and played loud music. After a long search for a simple cup of coffee, I settled in with a latte and a map in an espresso shop that was about to close for lack of business. Fun to have spotted a butterfly garden that actually had a monarch, yards from the deepest dip of the Snake on its whole long course; but it gave no directions. Where to go now?

Pooped, ending up in a rare motel, I fell asleep reading some lines of Edwin Way Teale's *Autumn Across America,* the third book in the great nature writer's Pulitzer Prize–winning quartet describing his travels through the American seasons with his wife, Nellie. I had virtually memorized the part where he wrote of "that double parade . . . monarch butterflies above and woolly bear caterpillars below . . . all across this rolling land." He called the monarch migration "one of the most puzzling features of the American Fall."

So it is. And nowhere does the puzzle beg resolution more than along the Snake River, a region where even Fred Urquhart — principal architect of the East–West monarch wall — suggested that it might break down. I would soon leave this great river that carries water from the summits of the Grand Tetons of Wyoming nearly to my front porch in maritime Washington. But before I did, I had a little detour to make.

NINE

*B*EAR

THE AUTUMN WAVE of monarchs in the West is a rarefied business. Instead of hewing to dramatic flyways, as eastern monarchs do, western migrants ride a broad front with occasional concentrations. While one such node accumulates in the lower Snake River Plain, I also knew that a substantial movement occurred along Utah's Wasatch Front, east of the Great Salt Lake. Years before, I had come upon an overnight aggregation on the Bear River Migratory Bird Refuge. I decided now to make a quick return visit. On the way out of Burley I stopped at the butterfly garden to meet Barbara and Richard Kerb and thank them for the monarch of the previous evening. I left with a bag of hot ginger-maple-raisin cookies.

I could avoid the interstate by taking a broadly parallel route along blue highways and dirt roads. After Declo and Albion, the way rose into piñon pines and aspens. In a high, spare valley I passed cowboys mounting for a roundup, and a tidy two-story red brick house with a sign that read "The Wood Ranch — since 1860." I passed through tiny Elba to tinier Almo, where false-fronted Tracy's General Store (1894) stood with a veranda over the mailbox and one gas pump. First built of logs, the store had an addition, with a dance hall above, made of adobe bricks kilned on the site. A handwritten sign on the wall advised "We have Jex Heaton's Bread 2# loaf, $2.25." I bought a cream soda, but I kicked myself for the next three weeks for failing to pick up a loaf of Jex's bread.

Down the road I stopped in at the headquarters of the City of Rocks Natural Preserve. White-bearded ranger Ned Jackson told me,

"We usually get monarchs moving through here toward the south, but I haven't seen 'em lately." From there the road wound west into a great granite jumble that looked like "a dismantled rock-built city of the Stone Age" to pioneer James Wilkins in 1849. It had been a busy place when the California and Salt Lake Alternate trails came through on their way to the Humboldt River across the deserts of Nevada: 52,000 people passed through in 1852.

Eroded, exfoliated, and scarred granite boulders made domes, peaks, horns, hoodoos, caves, arches, basins, and bowls — like a cross between Yosemite Valley and the Garden of the Gods. A mourning dove whooed out in the hollows. Yellow rabbitbrush blended into and surrounded yellow aspens, which shimmered bare or greenish, brass, red, and butter among the somber junipers and piñons. Pitons glittered in the granite along with mica and quartz, and exfoliated lenses, like the sheaths of an onion, left slits of blue sky between the remaining layers. The few people here now were mostly climbers. Two were belaying an impossible pinnacle nearby.

I don't like to watch rock climbers. They are all nuts, and I watch only from the same twisted compulsion that forces passersby to ogle a car wreck. I've skinned my shins on many a high rockslide, even seen a few summits. But gravity is not my friend. If I can't walk it, I'll leave it to the ravens and swifts. When the exhausted emigrants painted their names in axle grease on Camp Rock, none of them imagined skinny idiots in Day-Glo Lycra scaling the skyscraping rocks for fun a century and a half later.

Migrants of different species often find the same routes. I remembered once seeing black swifts and nighthawks crossing a high pass together at slightly different altitudes and directions. I supposed monarchs might have flown above the struggling wagons using the same passes, so I lay on orange-lichened boulders atop the northwest gap at the top of Emory Canyon, by the Bread Loaves (heavier even than Jex's), and looked up. I saw raven, magpie, eye floaters, flies, and fluff. I was hot in the sun, but the breeze had a sharp autumn edge to it. If monarchs passed overhead, they were way too high to see.

Almost endless rough dusty roads brought me to Naf, Idaho, as big as its name and as isolated as anything. Yet it had Pat's Mercantile Cafe and the big old clapboard Naf Dance Hall & Bar. I saw zero people in town, but the dance hall looked as if it was still in use. The cowboys come in and kick. Many times in decades past I had driven or hitchhiked along the interstate, miles away on the other side of a ridge toward Malta. I'd always wondered what was on this side of the ridge. Now I knew: Naf.

These economical names grown straight out of the sage (Jex, Naf) and the names transplanted here from immigrant pasts (Albion, Elba, Malta) got me thinking about the monarch's names. God knows I had miles to do it in. Its English monikers have not been many. Artist-naturalist John Abbot, whose exquisite painting in the extremely rare and valuable 1758 folio of his work is perhaps the species' earliest color portrait, called it "the Large Orange-and-Black Butterfly." "Milkweed butterfly" and "wanderer" have been employed now and then, the former especially in England. But for most of its written history, "monarch" has been used with remarkable consistency. Books often recount that the present name has something to do with the colors of William, Prince of Orange, stadtholder of Holland, and later king of England. Indeed, Don Davis, a wonderfully knowledgeable monarch aficionado in Toronto, told me that his late grandfather knew these butterflies as "King Billies." But the actual coinage of the term "monarch" may be attributed to Samuel Scudder, one of the greatest of the late-nineteenth-century lepidopterists. In a paper published in *Psyche* in 1877, "English Names for Butterflies," Scudder wrote, "9. Danaus plexippus. — The Monarch," and explained it thus: "D'Urban calls it the Storm Fritillary, but it is not a Fritillary. Gosse called it the Archippus, but this is not its proper name. It is one of the largest of our butterflies, and rules a vast domain." Next he named the viceroy.

The species' trail of scientific names has been much more tortured. The earliest known monarch specimen was collected in Maryland between 1698 and 1709 for the English naturalist James Petiver, who mounted it between sheets of mica. Thus preserved, it may still be found in the British Museum of Natural History, labeled in Petiver's hand and described in his catalogue with a long, prebinomial Latin phrase. The founder of modern taxonomy, Carolus Linnaeus, named the species *Papilio plexippus* in 1758. One Krzysztof Kluk erected the genus *Danaus* in an 1802 Polish paper. Through a chaotic nomenclatorial history that saw dozens of alternatives erected and torn down, even including at one time the viceroy's rightful specific epithet, *archippus, Danaus plexippus* prevailed and reigns today.

Butterfly names were frequently drawn from Greek and Roman history and mythology, usually with little relevance. Plexippus was the spiteful uncle of Meleager, lover of Atalanta. Danaus was the great-great-grandson of Zeus and Io, who ordered his fifty daughters (the Danaids) to marry, then kill, their fifty cousins; forty-nine complied and were condemned perpetually to carry water in leaky jugs in the lower world. In *Wings in the Meadow* Jo Brewer writes, "It is ironical that [they] should have chosen the names of two such miserable heroes for this remarkable insect . . . while the far more distinguished name *Parnassius apollo* was conferred upon a butterfly that in a number of ways is far more primitive."

I prefer to associate the name with Danae, the great-great-granddaughter of Danaus. Her father, Acrisius, fearing a Delphic prophecy that her son would kill him, locked Danae in a brass-bound hole. But Zeus came to her in a shower of gold and left her with child. The son born by Danae was Perseus, who would be the great-grandfather of Hercules and who killed not only Medusa, but of course, eventually, Acrisius. Gustav Klimt painted a beautiful pale nude portrait of Danae, Zeus's shower of gold falling about her prolific copper hair like butterfly scales, and that is the image of her that I carry.

Imagining monarchs among the golden aspens of the City of Rocks, I thought of Danae. And it is possible that Kluk did, too, when he named the genus. Latin grammar and later the Zoological Code prescribed that an adjectival species name agree in gender with the genus. Kluk was not obliged to make his new genus masculine to match the familiar species name, *plexippus,* but he may have done so for aesthetic reasons. Or maybe he confused the two and named the monarch's genus for Danae after all.

I carried with me, suspended from the rear-view mirror, a monarch fashioned from orange, black, and white-tinted feathers, which I named Danae. Sometimes when tagging a butterfly, I found that it pacified the insect to let it grip Danae while I worked on it. I then let it take off from the feather butterfly. Now Danae's wings flapped in the breeze as I sped over the first paved road I'd seen in many hours.

Out of the dry uplands, I entered fields of irrigated Mormon alfalfa and knew I was in Utah. The cut alfalfa was stacked in neat blocks of big rectangular bales like a giant child's play blocks or stored in little packets that resembled green ice cubes. As dark fell, I met the freeway again and pulled into Snowville for gasoline. A cold and vicious desert wind blew up and snatched my other icon, the wine-bottle monarch David Branch had given me that earlier windy night at Vantage on the Columbia. One second it dangled from the radio knob by its bottleneck loop; the next it was off to join the migration.

By the time I reached a rest area and information center on I-15 near Brigham City, it was way too late to visit the migratory bird refuge. Also, I needed a map, having lost Utah, so I decided to sleep there. Hiding from the big sodium vapor lights, I parked close up in the shade of a big camper. A little later it moved way down the row of parking spaces, perhaps because of the striped skunk that had been prowling its doorstep but wasn't bothering me. I lay back listening to a tape of Terry Tempest Williams reading *Refuge* and watched shooting stars.

She asked, Could migration be "an ancestral memory, an archetype that dreams birds thousands of miles to their homeland . . . not out of a genetic predisposition, but out of a desire for a shared vision of a species?" Terry was speaking of Canada geese, but it did not seem outrageous to me to imagine something that "dreamed" monarch butterflies thousands of miles to their safe winter harbors — if only Chance and Selection by other names. What is there, after all, besides memory and dreams, and the way they mix with land and air and water to make us all whole?

In the cool, lake-misty morning, the portly Canadian who owned the camper approached me to explain that it was the skunk that drove them away, not I. He handed me a bag of sweet cherry tomatoes from home, saying, "They're getting ripe too fast."

As I neared the Bear River Migratory Bird Refuge, barn swallows were massing by the thousands on wires and fences. There were four long, full rows of them. Terry had called this place "a fertile community, where the hope of each day rides on the backs of migratory birds." I heard "migratory butterflies," and my hope rode on their wings. Monarchs can be seen massed like those swallows in certain places, such as Point Pelee on Lake Erie, staging to cross great waters. I thought I saw one swallow that didn't fit into the crowd but dismissed it as imagination.

"Attention Swan Hunters" read a sign: "You May Not Possess or Use More Than 10 Shells a Day While Hunting Swans." Imagine hunting swans at all — and with rare trumpeters around! In Great Britain, all swans belong explicitly to the Crown, and no one may hunt them, except one venerable guild that may take ten per year. Uncharitably, I wished upon all swan hunters the remorse that visits the unhappy archer in the old English ballad "Pavonne," who in twilight shoots what he takes to be a swan, only to discover that it is his lover.

It had been nearing dusk when I watched monarchs going to roost here on October 19, 1980. Half a dozen or more migrants were taking their rest in Russian olives. I wondered if they would still make it, leaving so late in the year, the nights coming on very cold. Since then the Great Salt Lake had risen, topping the dike in 1983, and gone down again. With the flooding, as Williams describes it in *Refuge,* the place changed greatly. Now I found next to no brush and couldn't picture where I'd seen that evening cluster.

Driving the refuge road, what I thought were smoky plumes from hunters' fires turned out to be a vast hatch of minute brine flies rising in black puffs all along the dike. A western grebe slept, head tucked within its own white gyre that seemed mirrored in the coiled, darting necks of snowy egrets and avocets. A toast-brown hen harrier hunted and glided along the canal, and later, after many more females, one cool gray male harrier coasted low over the marsh. I'd looped the trapezoidal road bounding Unit 2 by one o'clock. The ducks were many but far out, spooked by early morning hunters; only the seldom-hunted coots ("mudhens") clung close to shore. *Zitta zitta,* said the marsh wrens; *whee* cried the avocets. The shy ibises were silent.

When explorer John Frémont came through in 1843, he wrote, "The whole scene was animated by waterfowl" that "made a noise like thunder." Today the nimbuses of tiny brine flies emitted a constant high hum. Monarchs, too, can be seen and heard in rustling clouds in their Mexican winter metropolis. I badly wanted to see at least one monarch here, but so far the only orange butterflies were alfalfa sulphurs and mylitta crescents sailing past the browning alkali bulrush and sago pondweed.

I turned back toward the Wasatch Front, streaked with the carotenes and anthocyanins of autumnal foliage among the blue-green scrub oaks and junipers. Distrusting my earlier judgment (or, rather, respecting my first impression), I looked again on the way out, and there it was, on the wires to the Bear River Duck Club, *one* tree swallow among thousands of barn swallows.

Following the lower reaches of the Bear River as it pours toward Great Salt Lake is a challenge, as most of the way is either fenced off or posted against entry by the refuge managers. I was approaching one of the few accessible sections along Corinne Road when a bright monarch crossed Powdermilk's hood. Heading due south, flying low and strong over a ditch, it passed over an expanse of dry teasel wands and blown pods of milkweed. Then it flapped into a stubble field, where I lost sight of it. I had a fast semi on my bumper, and by the time I could turn around and get back, there was no sign that a monarch had ever been there. I set out walking along the Bear.

A quarter-mile southwest I came to a canal paralleling the river, and a second monarch materialized, bearing south along the Bear River Road and ducking out of sight. It was lighter in color than the first one, and initially I thought it might be a big old viceroy from the ditchside willows. The flight manner, pattern, profile, strength, and direction all matched those of the obvious migrant, however, and it didn't return or stick around the willows as viceroys usually do.

Bear River itself originates in blue Bear Lake across the Wasatch to the northeast. It cuts north, slides around the top of the range, then swings south to consummate its brief course in the Great Salt Lake. Its delta, the Bear River marshes, once covered some fifty thousand acres. The much-abbreviated remains make up the refuge today. Many of the monarchs spawned east of the Snake River and south of the

Caribou Mountains are likely to funnel through the Bear watershed as it collects southward toward the Cache Valley, the narrow piedmont between Salt Lake and the western edge of the Rocky Mountains. Where this long plain, the heart, lungs, and head of the Mormon empire, is undeveloped and unsprayed, it generates many additional monarchs.

In Corinne, I called Ron Hellstern, a teacher in Hirum/Logan whose monarch-hunting students had found one hundred larvae in Cache Valley so far that fall. They had lost a lot to parasitic tachinid flies, but they had tagged and released more than thirty, the last ones just the previous day. The monarchs had taken off in various directions, but the four liberated yesterday had borne due south. The students' textbooks showed that monarchs from here should end up in California, but Ron told me that they had come to question this on their own. After all, one of Mary Henshall's Boise monarchs was recovered not far south of here, in Orem.

The tail end of the day took me down the big thumb of land that sticks south into the Great Salt Lake from its northern shore to the tip of Promontory Point. The highway beyond Corinne was puffy with bursting milkweed pods and jammed with traffic returning from the Morton Thiokol plant (of *Challenger* O-ring notoriety, as they hate to be reminded). But Promontory Road was deserted. The road was, unsurprisingly, washboard, but it passed through sunflower city—acres and acres of it, lush with yellow rabbitbrush as well. I found it hard to believe that these fields weren't swarming with nectar-hungry monarchs, and felt certain that a week or two earlier they had been.

The red-striped Promontory Mountains sheered up on my right, the Wasatch Front across the shallow water on my left. Five deer peered warily through sunflowers. At the end of the road, I found a trailer camp for the salt company, with causeways, railways, and electric lines all gone wonky and unstrung from the flood. I saw no humans, just dogs and mosquitoes. Half a dozen cottonwoods, one fixed with a swinging rope, stood between a dump and a graywater sump, the only trees visible for many miles. Orange butterflies were dashing about their crowns and basking, but as usual they were ladies.

Crystalline snowfields of salt ran out to ice-blue lagoons; ragged islands and mountains rose out of a salty fog. Thousands of phalaropes dotted the distant surface, spinning. I had floated in this sodium brew when I was five years old, and I could still feel the sting in my eyes as I tried to rub them on the skirt of my mother's bathing suit. Before sunset I took the road around to the west and checked a lone grove of tamarisk that should have embraced a cluster of monarchs but didn't. The road turned north, past a plain of otherworldly brine pans. From the air they look like paintbox color squares. Now their neon pastels refracted a heightened pink, purple, and gold sunset. Multiple mountain ranges stretched away in receding relief. Finally, the road died at a gate where a sign warned of cyanide cartridges put out for coyotes.

Heading back, I came across a road-killed "common" poorwill: so soft and such a huge gape, with its hypersensitive vibrissae around the surprisingly rubbery nose tubes, such elegant texture and striated pattern of browns. Then, just as Jupiter was rising, I hit an owl with the top of the car. I searched but could not find it either on or off the road, so maybe the blow was only glancing.

I remember most of the birds I've ever hit. There was a Hungarian partridge in Austria that I would have liked to eat. I struck a calliope hummingbird in Sequoia National Park on my way to give a ranger's campfire talk on "Birds of the Sierra," and took it along as a warm, tiny visual aid. There had been a western kingbird in eastern Colorado, virtually an omen of what was to be a bad afternoon, and a white-crowned sparrow (so pert in its death) on the way to Picabo, Idaho, in the snow. The worst was when I saw a twinned blur before me in the dusk, felt a double thump, and recovered the ruffled bodies of a Swainson's thrush and the screech owl that had been about to take it.

As I recited the roster of all the poor birds I'd killed, a suicidal jackrabbit shot into the road. I hit the brakes hard, fishtailing in the dust. The big-eared hare actually dove under Powdermilk, then seemed to bounce right out again. I missed it.

The long, bumpy track merged with the blessedly smooth Cedar

Springs Road near the Golden Spike Memorial, where tracks from the East and the West met in 1869 to make the first transcontinental railroad a reality. From here I could have simply continued south through Utah in the direction of the afternoon's monarchs, and surely found more to follow. But just as I had rejected the expedient route south from John Day to the lakes on the Oregon/California frontier, I felt it would not be right now to forsake the essential course I had been shown from the start. I was satisfied that the wanderers moved south through the Cache Valley. Now I wanted to see what happened at the rim of the Snake River Plain, where monarchs must strike up, out, and into the great beyond.

The diagonal of Highway 83 on the way back to the freeway took me past Morton Thiokol. The plant was an entire city of yellow lights that went on for miles: blocky buildings with a *giant* American flag and a huge "SAFETY PRIORITY 1" sign, jaundiced in the sodium light. I pulled in and quickly rounded a rocket display at the entrance checkpoint. It struck me that this must be where all the rocket scientists it doesn't take to do this and that are to be found. I couldn't help thinking of the very different research of one aeronautical scientist I know.

David Gibo of Toronto uses gliders and radar to study how monarchs migrate by flapping, gliding, and soaring. He believes that they can adjust their wing loading (ratio of weight to wing area) by taking on or expelling water as their lipid masses wax or wane with feeding and age. Through differential ballasting, they can actually adjust and maximize their glide angle. Good gliding is vital to migration: earlier, Gibo showed that monarchs could glide for more than a thousand hours on a 140-mg reserve of fat, but go only forty-four hours on the same energy if they were cruising and flapping. Unless they can glide much of the way, they can't possibly make it. So any adaptation that favors gliding will also enable a butterfly to fly long distances in the time available while still pausing at length to nectar and sit out bad weather.

Gibo first showed that for stable gliding flight, the monarch's center of gravity has to be located slightly behind the midthorax. Since migrating monarchs take on considerable weight in the form of lipids, which are deposited in the abdomen, they can stall in flight unless that weight is compensated for. By taking on water ballast in their crops, monarchs may increase wing loading and air speed without changing the optimal glide angle (about 3.5 meters of forward movement for each meter of sinking). Without water ballast, they have to alter their wing placement, resulting in slower flight and faster sinking. Gibo tested this hypothesis, measuring lipids in earlier- and later-phase migrants, and found strong evidence that they did use water ballasting to offset fat accumulation and maintain the glide angle. I doubted I could see anything behind Thiokol's yellow lights more impressive than that.

On long drives, deep in the western night, I listen to the radio with a perverse affection. The music fades in and out, heavy on country and oldies but ranging from Perry Como to Bob Marley. The talk shows, with their interminable advertisements for various nostrums "not available in stores," run from the infraright to the ultraleft. Now I picked up classical from Newfoundland, now bluegrass from Las Vegas. Trolling for a little public radio near the bottom of the dial, I perked up at a familiar voice. It was Katrin Snow, once a monarch butterfly conservation coordinator for the Xerces Society. Now here she was as a radio announcer, broadcasting into the Utah night between Doris Day and Dolly Parton. There's no telling where a fling with these bugs might get you, I thought, as two coyotes sitting by the roadside welcomed me back into Idaho.

Well after midnight, I settled into the day-use area at Three Island Crossing State Park, on the Snake River between Twin Falls and Boise. The campground was full, but I reckoned it was near enough sunrise to call my stay "day use." Half a moon rose left of Orion, and a great horned owl hooted outside my open window. A second owl answered from way off to the right, then came closer. A coyote shrilled on the far left, as if in chorus. The second owl moved right in, and the two boomed together. A car checked me out; either it was another night traveler, or the ranger felt charitable.

Closer to dawn, the great horneds moved off, and a barn owl screeted one time. The night was almost balmy, and the smell on the air was of the river. From Three Island Crossing I would head toward Adobe Pass, Nevada, and the Humboldt River through Duck Valley. A line drawn from Hell's Canyon through the canal where I'd seen the monarch and the otters would point this way. It seemed as good a path as any for trailing southbound migrants out of the great sink of the southern Snake.

The river ran brilliant with sunshine, asters, and whiskey-colored cattails. Greater and lesser yellowlegs migrated alongside a white-fronted goose, its mate absent, whose lonely two-note call echoed over the smooth water. The resident kingfisher, a red-belted female, rattled willow to willow as if impatient about staying behind. I felt an odd reluctance to leave this river I'd run with for so long. Then the reason struck me: the Snake runs into the Columbia, the Columbia to home. When I crossed the looming divide into the Great Basin, I'd be out of my own watershed.

I hadn't driven far before I stopped to walk along a riverbank road lined with shad scale and willow. The flowery verge was alive with butterflies — common and orange sulphurs, western and Becker's whites, crescents and coppers, ringlets and skippers. West coast ladies had been a frequent feature of the trip, but now the first *Vanessa cardui* flapped past — the basic painted lady. It is also called the thistle butterfly, for its favored host plant, and the cosmopolitan butterfly, because it occurs on every continent except Antarctica. The small and tatty spring immigrants from warm latitudes beget robust, bright salmon-colored grandchildren such as this by October.

As Highway 51 rose from the irrigated bottomlands toward pygmy steppe and crested a road-cut rise, I saw a big shiny monarch nectaring on a lone yellow rabbitbrush. It was fresh; I could not tell its sex. It lifted off before I got a stroke, and as it hovered, I was reluctant to take a risky wing shot. Instead of alighting again for another sip, it fluttered off uphill and flew lazily across harsh red slate. A fire had passed this way not long before, and the monarch looked like the last live flame licking over the burnt slope. Then, rising toward the ridge, it picked up speed, pointed sharply SE, and vanished.

Grasmere was only a couple of shacks and a bar among the sagebrush, a world away from Wordsworth and Beatrix Potter's Grasmere in the English Lake District. The Idaho town's founder must have been feeling either wry or homesick. Was his neighbor in the next wide spot wondering why he'd come when he dubbed his barren plat Riddle? Between the two ghost towns, a shockingly beautiful aspen grove erupted from the desert, the biggest tree so robust and brown-trunked I first took it for a cottonwood. Rabbitbrush and big basin sage bloomed yellow all around the half-gold, half-green clump of aspen. Old trees where a spring once flowed — a classic roost, where generations of monarchs may have gathered, perhaps as recently as last night.

On the Duck Valley Reservation of the Shoshone Paiute, stalls made of pole frames thatched with willows stood in what looked like a sometime market or powwow ground. Magpies took shade beneath them, and in a loop of the blue stream I saw the ducks of Duck Valley: mallards. The tribal center was in Owyhee, where handsome old stone buildings stood derelict and an old yellow dump truck had been a planter for years, but the high school — "Home of the Braves" — was new. Nevada began; here was the first casino.

From there the butter yellow of the willow valley climbed into the lemon yellow of the aspen hills. At Mountain City the slopes of Humboldt National Forest ran to sage with velvet pastures in the bottoms. "Sunflower Flats 11 miles," up a side road, tempted me, as did a very inviting campsite where Wild Horse Creek met the Owyhee River. But it was already five o'clock, Pacific Time again, and I was eager to put the summit behind me. Upcanyon, a double-curvature, thin-arch, ninety-foot-long dam, built by the Bureau of Indian Affairs and the Bureau of Reclamation, choked the poor little Owyhee. The 71,600 acre-foot Wild Horse Reservoir was "Named for Wild Horses that Roamed the Area." Past tense.

Driving on, I felt something crawling on my foot and saw that it was a big, black, orange-spotted carrion beetle — it must have landed on me when I moved a well-infested badger off the road earlier in the day. Carrion beetles reek appallingly of putrefaction, and who knows what they might carry, rooting around in rotten meat as is their habit.

I pulled off the road to remove it, but then I couldn't find the stinker. I was afraid it might be with me for the duration, burrowing into my gear, my food box, or my sleeping bag.

While I was stopped, out on a high, open place, another monarch flew directly and closely overhead. It was beating strongly SE, heading for Adobe Pass. I followed it with my field glasses for a few hundred yards as it continued on an unvarying trajectory through the still air until it receded to a point, then nothing. So, thanks to another orange-and-black distasteful organism with whom I was not eager to share my personal space, I was granted one more vision of what had brought me here. I ended the day, and the Pacific Northwest, with good solid tracking up the Owyhee headwaters and into the territory of the Humboldt. I raced the sunset to Adobe Summit, and it won. Crossing over, I entered the Great Basin.

TEN

Bonneville

When I had imagined roaming the basins and ranges, I'd pictured dramatic and lonely landscapes scratched with rattlesnake draws, rough bivouacs in shad-scale thickets, coyotes calling to stars undimmed by human lights. Yet here I was again, hunkered in my car in a noisy, lit-up freeway rest area. I had my reasons, and they would turn out to be remarkably good ones.

From Adobe Pass I had dropped down into Elko, Nevada. I took a walk and donated eight bits to bandits that made electronic beeps instead of satisfying clunks and didn't even have cherries. On some Nevada crossings, that much invested has won me a meal and a tank of gas. No such luck this time. I dowsed some local ale and a decent bite out of the neon streets, and over the second pint, it was time to decide: turn west along the Humboldt River to where it sinks into the playas of Ancient Lake Lahontan, then presumably up Donner Pass and down to the coast? Or continue southward?

Well, dammit, I hadn't seen a single southwesterer for days. Those butterflies had been heading *southeast* as they rose off the Snake Plain. So I headed out of town east on the interstate to intersect their course. Putting Nevada quickly behind me, I passed through the border town of Wendover, where thousands of trucks were lined up at the casinos, whose billboard promised "17 acres of parking." All those truckers and gamblers, all the night riders across the endless alkali strip of I-80, all looking for luck . . . and I, alone in the entire West, had the temerity, the presumption, and the sheer *luck* to be following monarch butterflies. I also had a good tip.

Once I'd settled into the freeway rest area on the Bonneville Salt Flats that night, I leaned back, turned on the overhead, and reread the quotation by Edwin Way Teale that had drawn me there. It was from *Autumn Across America:*

> Before us now the Brown Mountains of the Pilot Range rose higher at the western edge of the desert. But we were still well out on this dead, flat, shimmering land when we encountered one of the most amazing sights of our travels.
>
> Three monarch butterflies drifted by, then two more appeared in sight — five monarchs winging their way across the desert on their long migration south. More than fifty barren miles of salt and alkali lay between them and the southern boundary of this desolate, plantless waste. How, amid heat wave and mirage and blinding glare, could these insect voyagers find their way? They were not using landmarks, for there were no landmarks to use. All was one white, level, featureless plain. They were not following the terrain — the lay of the land, as in a river valley — for the desert stretched before them to the horizon as level as the sea. How were they orienting themselves?

Although my two vanishing bearings from yesterday's monarchs pointed here, speaking strictly statistically, my chances of duplicating Ed and Nellie's lucky sighting seemed vanishingly small. But walking back from the head later that night, it occurred to me that the trees of the rest area — fourteen robust Russian olives in seven watered planters — comprised an artificial oasis that was an obvious monarch roost. In the midnight glare of the sodium vapor lights, I canvassed the trees from the ground and from an observation platform, but saw no obvious clusters.

I fell asleep remembering the last time I'd been here, with Thea, Tom, and Dory. It was the year of the Great Salt Lake's apogee, and the freeway, the rail lines, and the Saltair Pavilion were all marinating in the risen lake's brine. That morning we had watched an anise swallowtail emerge from its chrysalis on the front steps of a casino in Wendover, and the next day we had released a freshly hatched Nokomis fritillary on pink cleome near Vernal, on the other side of Utah.

In between, we'd stopped at this same rest area, climbed the spiral ramp of the observation platform, and gazed across the shimmering clean sheet of the playa. Shading my eyes from its white-hot glare those years ago, I never even imagined the possibility of monarchs. Now I did, thanks to Teale, but the odds seemed very long indeed.

I get up when the sun hits the trees, meagerly — this Sunday being the first hazy day in many, with a cool, very light breeze off the eastern desert. Crawling from the car, I take up a position on the salt flat in full view of the linear grove (and of its twin, twelve trees in the westbound pull-off across the freeway). If there are any monarchs here, I should see them rise. The unutterably flat, Tide-white surface reaches away toward the Bonneville Speedway, where people come to test jet-driven cars against the restraints of gravity, friction, and drag. The saline, bleached pallor of the place seeps into my eyes and threatens to erase everything else, everything that is not white and horizontal. I feel vertiginous.

I watch for an hour, from half past eight to half past nine, standing still. I am beginning to think about breakfast when suddenly a monarch lifts off from the nearest and largest pair of trees. I do not actually see it launch from the branch, but it appears at tree level, flying slowly, just getting going, and it must have just risen from the boughs. The temperature is 63 F., maybe warmer on the east side of the tree; the awakening conditions much like those at Crab Creek, but less windy. In fact, the air is nearly still. But all of this I recognize afterward. For the moment I am transfixed by the rising butterfly.

I don't know if there are any more, as I'm not about to take my eyes off this one. The monarch circles and flutters, gaining a little altitude, just a few feet away from me. I don't want to catch it, just watch. It drifts west a little on the light easterly breeze, then glides around in spirals, gaining height. At fifty feet or so it glide-circles counterclockwise past the sun, and its backlit wings become chrysophanous panels of stained glass. It is basking on the wing, as butterflies often bask on the ground, gaining warmth, still sluggish.

A barn swallow appears, and I briefly fear for the rising butterfly. Barn swallows, ring-billed gulls, and English sparrows are the only birds here, and all are looking for breakfast. The hirundine makes one investigative pass at the butterfly, the monarch flies momentarily back *at* it, and the swallow veers away. *Danaus* continues to spiral for a minute or two longer. Then, just three minutes after arising, it begins to flap hard and fast, making forward as well as upward progress to the south. Once more it stops at 100 feet or so to spiral, and again higher; then, at maybe 150 or 200 feet, it takes off at a moderate clip — flap-flap-flap-flap-glide, flap-flap-flap-flap-flap-glide — heading SE. (When I later describe this to David Gibo, the monarch aeronautics authority, he says, "That butterfly knew where it was going!")

The monarch passes over a long Union Pacific train chugging east. I watch for about a mile before it recedes from sight. Then I mark its trajectory, which is steady for most of the way, as it vanishes ESE at 120 degrees. The breeze is very light from the east. When I rub my eyes, they are damp from the white sun, the salt, and the good grace of chance.

A whiskered and bellied truck driver watched me taking the monarch's bearings. Later he sidled up, observed me cooking oatmeal on a picnic table, and said, "What's that supposed to be?"

"Oatmeal," I said.

"How do you make it?" he asked.

"Put oats in water."

"Huh." He ambled off. Soon back, he said, "Those salt flats are pretty nice."

I replied, "Yeah, you can see why they like to drive on them. But when they're wet, you'd probably break through somewhere."

"Yeah," he said, "sink right in," and walked away. He didn't ask about my lines of stones and compass. I felt a little bad about not being more forthcoming. But while the shivers of this experience were still reverberating, I wanted to write up my notes, not gab. I wanted to conjure on the monarch's course across the flats, the continent. Nearby a Goodyear-sponsored interpretive sign gave the land-speed records for the Bonneville Speedway. By comparison with what I'd seen, they seemed no big deal.

I felt great satisfaction that the oasis idea had worked. True, it was only one butterfly, but it was a very good start for the Great Basin. Before I left the rest area I gave my accumulated cans to an elderly Japanese-American couple who were gleaning aluminum from dumpsters and barrels. They seemed politely but not abjectly grateful. As I pulled out, the old man sat in his old car and smoked, looking tired and worried, as his wife continued to search. I thought once more of Teale's impressions of Bonneville. "It is easier," he wrote, "to accept the message of the stars than the message of the salt desert. The stars speak of man's insignificance in the long eternity of time; the desert speaks of his insignificance right now."

To go southeast I first had to drive east for a while, since everything south for many miles was off-limits, given over to the top-secret weaponry of the Wendover Range, Deseret Test Center, and Dugway Proving Grounds. My Bonneville monarch would be crossing them by now, oblivious to signs and fences, but I was obliged to go around on the Dwight D. Eisenhower Highway. At thirty-eight miles, I took an exit called Knolls and turned back west on a remnant stretch of the old two-lane blacktop the Teales had traveled some forty years earlier. A relict U.S. 40 sign pointed to "Wendover west, Salt Lake City east," the white of its letters wind-blasted off the pitted and faded green.

Driving, I was hypnotized by white on my right, a coal train flickering on my left. I had to remind myself to stop before the blacktop ran out. *Salicornia,* a succulent salt-loving plant I associate with the shore, grew beside fluffy yellow heads of rabbitbrush crowded with thousands of pale millers and grass moths. Sparrows and horned larks were flycatching and hopping up off the road to get at them. I walked barefoot out onto the salt flat. At the rest area the salt had been cold and very sharp. Here it was finer grained, cool, and soft like a sponge. The dried brine was raised in bumps, warts, and folds that gave to the touch the way a glazed doughnut does, but sparkled like a sugared doughnut. I walked until the salt began to sting in the cracks of my feet. Ants and deer had left the only other tracks here in the bed of Ancient Lake Bonneville. I saw no more monarchs.

I crossed the Cedar Mountains Wild Horse Range at Hastings Pass. The Donner Party came this way prior to their bad time in the Sierra. Some believe that Great Basin monarchs reach the coast *via* Donner

Pass, and perhaps some do; but to follow that route, the ones I'd seen would have to take a sharp right turn somewhere. At last, at the bottom of the Great Salt Lake, I was allowed to make a right turn myself. I was almost due south of and only about thirty-five miles from Promontory Point, having nearly closed an enormous circle in three days. I hadn't expected to come so far east again, but tell the monarchs about it. The Stansbury Mountains would keep me well insulated from the urban corridor I had hoped to avoid. But it looked as if the paths of the Snake and Bear monarchs would merge after all, somewhere southeast of Bonneville.

In midafternoon I came to Horseshoe Springs, a BLM wildlife area, nectarless though fenced from cattle. The springs, arrayed in a big C, were roofed with algae and pondweed in long green tresses. The day was up to 80, and in a cool, open pool I washed the stinging salt from my feet, tickled by a little bass. Tall mountains on both sides of a broad valley overshadowed blue-green pools reflecting blue-cirrus skies; hundreds of bluet damsels made a looping frenzy over the water, and large orange dragons gave me a start.

On the Skull Valley Gosiute Indian reservation, the shiny new Pony Express store was open. The proprietor, a middle-aged Indian man, spoke quietly of monarchs. "Oh, yeah, I know them," he said. "Big orange ones. They should be coming through." The route of the old Pony Express ran nearby. Its dirt ruts would take me southwest toward Fish Springs National Wildlife Refuge, but I took the paved road south instead. It ended at a formidable guard station for the Dugway Proving Grounds, where what is proved you don't want to know. A lepidopterist friend of mine, Karölis Bagdonas, had been sampling Dugway as part of a major ecological survey of Department of Defense lands in which he was involved. He had seen amazing ecosystems, a number of monarchs passing through, and incredibly grim pollution. The genial guard said "the professor" had left.

Outside the gate loomed a huge Mormon church in Early American style, white fenced and equally forbidding. Turning there, I passed a place called Terra tucked into the junipers before hopping over the Onaqui Mountains at the south end of the Stansburys. The handsome Rush Valley lay below the Oquirrh Mountains, beyond the village of

Ophir. Indian and Old Testament names merged in a local vernacular that sounded almost in concord, though their histories were anything but. Later I would pass Fort Deseret, built by settlers in eight days in 1865 to protect against the Pahvant Indian uprising. The Deseret Test Center, west of Dugway, echoed the name.

The Book of Mormon tells the story of the Jaredites, one of the "lost tribes" that reputedly fetched up in the Americas. The book states, "And they did also carry with them deseret, which by interpretation, is a honey bee; and thus they did carry with them swarms of bees, and all manner of that which was upon the face of the land, seeds of every kind." Brion Zion, a self-styled spokesman for the canyonlands, later told me, "It was in tribute to the Jaredites' cooperative, communal quest for spiritual freedom that Brigham Young gave the name Deseret to the Latter-Day Saints' political and theocratic kingdom in the Great Basin."

Utah is still called the Beehive State. Since the name Deseret stems from the pacific Mormon occupation of beekeeping, its repeated use for military sites struck me as odd. But fortified establishments have long been part of the Utah scene. Now I passed a range of earth-covered bunkers resembling the nerve gas and munitions storage mounds at Hermiston, Oregon, much more than the sugar beet houses near Shoshone on the Snake. This peaceful landscape hosted an astonishing array of warlike purposes.

Beyond the Faust Station of the Pony Express stood a baker's dozen of pronghorns, like echoes of Africa, grazing out in the greasewood. Before day's end I scanned a prominent clump of juniper and piñon — a likely roosting site for monarchs, like the pine and cypress mix of Monterey. But sunset came at Tintic, Utah, in the absence of monarchs. The day went out on three dispiriting notes.

First, the Tintic Valley Cooperative Research Project, a 4,500-acre tract set aside in 1949 by the BLM and Utah State University "to study responses of livestock and vegetation to various management practices." The practices: controlled grazing, sagebrush and juniper eradication, and grass reseeding. The responses: increased grazing capacity, better quality forage, increased livestock gains. What a surprise.

Second, the White Sand Dunes. I'd planned to camp there, but found legions of ATV-hauling trucks rumbling out at the end of a dune-busting day. Though biologically and scenically unique, the dunes are described in the BLM promo brochure as "Little Sahara Recreation Area: Utah's sandsational OHV Playground." The flyer goes on: "Welcome! Wondering where you can power a hill and glide . . . or bounce yourself silly amidst the boonies? A place where you can hang with the crowd or saddle up and ride off into a sage-tinged sunset: Then look no further than Little Sahara. It's your Great Basin ride-a-way in the Sevier Desert of central Utah. It's a place big enough and diverse enough that you'll want to discover it . . . again . . . and again."

Not actually looking to bounce myself silly, I pressed on to Delta and took a minimalist motel room from the aldehyde-breathed manager. The final downer was the sole mention of the land in the presidential debate, which I watched on television: "Protect the environment while growing the economy." Road-weary, cow-sour, and dune-sick, I crawled into the bath and on into bed. The morning had been great, and I was still tingling from that magical monarch, appearing out of the silver-gray leaves and disappearing across the diamond-white flats. Monarchs, after all, are comparatively rare in the West, and I couldn't complain. But I had hoped for more.

In 1540, Francisco Vásquez de Coronado, governor of Galicia in New Spain, led some three hundred conquistadors north in search of a chimera and a dream: the Seven Cities of Cibola, and Eldorado. The fabulous Cities of Gold, reported by Fray Marcos de Niza and reaffirmed by Mechior Díaz, naturally exerted their lure on the gold-hungry Spanish legions. Coronado roamed over hundreds of miles of the Great Basin, finding the Zuni Pueblos instead of any gilded Cibola, and a village of Wichita tepees in Kansas where Eldorado was supposed to lie. Finally, perhaps understanding he had been gulled by greed and fancy into chasing a mirage, he retreated to Mexico in 1542.

When I say I'd hoped for more, I did not anticipate some sort of butterfly Cibola, masses of golden wings secreted in the folds of the Utah deserts. I never expected Eldorado. But more, perhaps, than I had found. Now, lying in the Delta motel, I didn't know when the wishful conjuring left off and the dream began.

The western edge of vision has the fiery intensity of an Arizona Highways *sunset, but that isn't what I see. The sun bores straight and narrow through a hole in dull clouds that rub the horizon free of color. The oranges and reds come entirely from the backlit wings of butterflies: four times as many gold-foil panes as the hundreds of monarchs clinging to the tamarisk at the top of the wash, where I have followed them to their evening's rest. Their soft cluster-flutter keeps up until the dusk turns cool enough for wool, for fire, for stillness. Then the brilliance fades to a vague peach glow in the brush of the bivouac. Soon the desert stars take over even more, even brighter than the butterflies themselves.*

When I sleep, cocooned just yards away from the hunkered bunch, there is nothing to tell me the monarchs are here beside me save memory and faith. I wake and turn as the moon rises full. In its werelight, I think I can see dull color in the cluster, a smudge on the night sky like bruised persimmons: enough to make me think it's real.

And when the cold sunrise again picks out their hue from the wispy olive blur, the low beams, raking straight across instead of burning in from behind, daub the mass more nutmeg than cinnamon, more maple than marmalade. As some of the venturers spread their wings to capture the early rays, their massed palette grows brighter by measures, then takes on an oriole intensity I've never seen before.

The shivering begins — first one, then several, then all together — until the entire tamarisk looks ready to take off. A magic degree is reached . . . the first great glider rises . . . and one by one, all the monarchs launch into another day.

This, then, is how exodus goes. You get through the night as best you can, shiver against the chill, accept what the next day has to offer as if you had a choice, and then — just go! I do the same, and I try to follow. But I haven't even stuffed my sleeping bag, and already the monarchs are sailing over the far flat, sliding across the endless desert, sideways to the rising sun.

I blink, and they are already gone.

In a bleak land of shadscale and salt pans, who could resist a sign that reads "Sunstone Knoll"? It was a private gemstone claim. A twenty-five-foot-deep pit had been sunk at the base of the knoll, a shallower one on top. Peering in, I wondered what the "sunstone" of the knoll's name could be. Guys in the café had been talking about topaz. A sagebrush fence lizard, cryptic against the burnt sienna rock on which it was sunning, allowed me a close look, and in return I did not pull its tail off. Down below on the flat, two yellow engines tugged eighty-four hopper-cars filled with coal from a nearby mine. The engineer hooted to me on the knoll. For an instant I could see the attraction of coal mining or rockhounding, the pull of a quarry that doesn't move. But gems were the domain of the folks who had signed the register in a tin can near the pit. I pocketed some pretty rocks and moved deeper into my particular haystack — for now, the Great Basin. Gemstones or butterflies, the plan is the same. What you do is, you jump right in and hope something pricks pleasantly. For, as one friend reminded me, some haystacks *have* no needles.

I turned off at noon to the Clear Lake Waterfowl Management Area. Out in the middle of the Great Basin, the flora was largely a mystery to me. Little red halophytes grew on the floor like bits of leather, and I had no idea what family they belonged to. It was like visiting a town where you know no one. Then you see a half-familiar face, and the world comes into focus again. Here, for example, a little canal was nicely lined with milkweed, long gone crispy. Most of the local monarchs were probably long gone, too, so any I might see would likely be emigrants from farther north, like myself. I heard a rattle among the milkweed, and a little gopher snake slithered. Gopher snakes mimic rattlers in appearance and behavior, even imitating the buzzworm's alarm, but a glance at their heads reveals the absence of viperhood. It paused at my tread.

Inside the gate of the wildlife area, a white, Boot Hill–type wooden tombstone was inscribed: "To the / memory of / curious little / hands / 1949–1954 / I'm here / because / Dad left / loaded / guns / lying around." A small femur was wired to it — good God, I thought, is it the kid's? The bone was planted in a child's shoe, almost rotted away, with a plaster foot; the other foot was there too. Farther on, another tombstone read: "To the memory of a fool who drove with loaded guns." I decided these grim warnings were black humor; but they got me at first. And what about the specific dates? I'm still not certain.

Needful of a long walk to work off the road food and butt kinks, I strode out onto the rushy paths with Marsha. The north marsh had been burned to a matte-black filigree and had that dead, clean smell of a recent grass fire. Two ravens rose over it, black on black; a red-winged blackbird's epaulettes glowed like surviving embers. A mountain chickadee, with its white eye-stripe, popped up in a saltbush; mountains rose all around, none of them near. Less surprising, a sage thrasher, with its scythelike bill. The diked lakes were devoid of waterfowl except for coots, which walked on water, then fell through, inching away from me. Just once they flew a few feet, all together, with a roar like a jet, spooked by Marsha. And a few mallards, who spook at anything, and who can blame them? Wildlife refuges are hunting places, after all, and this was the season.

Harriers abounded, and a male made like a falcon and swooped down on an aerial flock of blackbirds, missing. It reminded me of watching buteos dive on bats the previous spring in Texas. The Mexican free-tailed bats were flowing out of a cave in the hill country of Texas by the millions, looking like the smoke plumes of midges at the Bear River Refuge, but thicker. The redtails and other soaring hawks weren't particularly adapted to bat hunting, but in the presence of so much food, how could they resist trying? Occasionally one snagged a nice big pregnant bat almost by accident. Animals do make decisions outside their normal instinctual patterns. A monarch makes a kind of "decision" when it drops from the sky to seek shelter or deviates from its heading to work a roughly parallel river course for nectar.

Instead of going back the way I came, I followed the causeway, hoping it circled back, and it did. The landscape came in bands of color: gray road, pale green saltgrass, yellow milkweed, pale amber cattails, deep amber rushes, blue water, green far slope, brown basin, tan and mauve mountains, pale blue sky.

In the field alone I often talk softly to myself. Now, rounding south to the road, I said out loud, "The rabbitbrush back toward Powdermilk might be productive." On the exhalation of the last syllable, on the first bright rabbitbrush, I saw the flicker of Big Orange: a huge male monarch. It lifted off the plant, oblivious to gravity, and flitted from one *Chrysothamnus* to another, investigating blooms that had largely gone past. Finding a fresh one, he settled. I approached, but before I could make a swing he popped up again and began to rise. Like yesterday morning's monarch, he climbed in small circles and spiraled up to three hundred feet or more. Then he powered off SSE at 167 degrees for about a mile. But instead of vanishing into the zenith, this one volplaned down to another big clump of rabbitbrush in the distance. This was more like it. In hopes of finding him again, I followed his trail in that direction.

As Stephen Trimble says of the Great Basin in *The Sagebrush Ocean*, "The more we know the more we see. Attentiveness counts for everything." I was attending, all right, but as a near-blind foreigner or, at best, through the compound eyes of a visitor. And what does the monarch see through those compound eyes? As he climbed, sailed, and dropped, what view did he obtain? His eyes consist of thousands of hexagonal lenses, each connected to the optic nerve by an individual receptor unit called an ommatidium, and each responsible for a degree or so of the visual surroundings. The image collected is a single, integrated scene, not multiple, as is often assumed; but it is also distorted by the curvature of the eye, and its acuity is estimated to be about one percent as sharp as our own. However, butterflies' "flicker fusion frequency" is much greater than ours, so they can perceive motion extremely well, and their field of vision is much broader, taking in well over 300 degrees of the scene around them. As for their color vision, they perceive the spectrum from ultraviolet deep into the red, as wide a range as honeybees and somewhat wider than our own. They utilize the ultraviolet extensively and may not see colors the same way we do because of UV fluorescence.

So when I approached this Clear Lake monarch, though Marsha and I were only vague forms, he detected even my slow, fluid movement as a series of smooth, alarming flickers. Some small motion I made put him to flight. The blue sky and the brown land contrasted strongly. And once the monarch was up, even on high, stripes of green, meaning moisture and shelter, and patches of yellow, shouting nectar, would stand out for him beautifully. Swallows swooping nearby would look like Day-Glo sparklers, and a fritillary flying near might draw the same initial response as another monarch. Once the insect was down again, it would be no trick to find a flower or a perching leaf, and he might be able to make out a spider web in the sun, but probably not a cryptic ambush bug or mantid. In other words, my monarch could see much better than many have thought. Not by a long stretch is he flying blind.

The Black Rock Desert was black, all right. I crossed deposits that I first thought were slag cinders or asphalt loads discarded on the ancient lakebed. But it was black pumice and pudding pahoehoe, lava that looked a lot like double-dark chocolate fudge abandoned in midswirl. This wasn't the glistening floor of the Black Rock Desert that Barry Lopez and Gary Snyder have written about; that lies way off to the northwest, in western Nevada. But this one was as swarthy as the Bonneville Salt Flats were chalky. In places you could walk on fine black gravel leading, like a raked Zen garden, to pockets of little pink succulents sprouting among the ebony pebbles, to vermiform desert driftwood enclosing pocked lava lamps where silicate solutes had evolved into the conical teeth of extinct saurians. With jokes like this lying about, it is no wonder good Mormon children believe the fossil record is a prank God plays on the gullible and those of little faith.

In the direction the monarch had taken, a rampart of columnar-pillar basalt all along the north side of the road gave way to fields and great stacks of hay bales, in total contrast to the jumbled lava beds along the south side. More color fields: straw and aster ditch, bright green alfalfa, lemony rabbitbrush, shiny black lava, mountain-bluebird sky. I parked, crossed a field, and clambered up a swale into and across the treacherous lava. It clinkered and tinkled underfoot, *very* sharp — don't fall! The edges were glassy, breakable, and bubbly. The flow seemed relatively recent but carried loads of lichens and, surprisingly, moss in the deeper clefts. Raptor down and spider strands caught on the ragged edges, waving gossamer. I dropped again to the soft benignity of the alfalfa, and cruised the chromium-yellow border of the flow. The orange flickers I saw were too big (flickers) or too small (a possessive crescent, a heavy-drinking Vanessa). A red admiral shot across the ragged surface like something spewed from a hot vent out of the magma below. Blues, coppers, and sulphurs flashed back and forth, and many moths; a junco jumped up and got one. But no monarch. If the last one I'd seen had settled here, it was already gone again.

My plan had always been to orient myself according to the most recent monarch's vanishing bearing, then, locating the desirable resources along the way, to find another. To an extent this was working, as it had at Clear Lake. But now, scanning the yellow band of rabbitbrush, I realized that this particular nectar source was so ubiquitous across the basin and range region as to be less than precise in locating the migrants. Rabbitbrush is considered by agronomists to be an "increaser" on heavily grazed, degraded rangeland. Just as cheatgrass has largely replaced native bunchgrasses in the West, rabbitbrush has smeared across the range of other native nectar plants. With more milkweed and more concentrated nectar, the monarch migration might historically have been larger and less dispersed in the West. Rabbitbrush was good for the monarchs and fairly good for finding them, if one was willing to survey acres and acres of it, but hardly the honeypot I'd imagined. It was a buyer's market out there for the danaiine nectarers.

Hoping to tease a buyer out of the sunflower verge, and to avoid I-15 as long as I could, I cruised an intensely hot frontage road. In Meadow and Kanosh the sharp, thin spikes of LDS churches poked complacently between alfalfa cut in neat rows. Finally I had to merge with the freeway. Francisco Atanasio Dominguez and Silvestre Vélez de Escalante had come this way in 1776, and in 1843 came Frémont, who was the first to call this vast drainless bowl the Great Basin. In return, his name was attached to the golden cottonwoods hereabouts, to a river farther south, to several counties in adjacent states, and to the early occupants of the region. The so-called Frémont people occupied a large chunk of present-day Utah, where they grew corn, squash, and beans from around A.D. 400 to 1350, a period whose moister climate permitted agriculture. Now the farming was irrigated, the Frémont long gone. At a shady rest stop, where timid piñon jays flashed turquoise in the trees, an Indian woman sold jewelry plucked from their feathers to a pigtailed truck driver.

Over much of my day's travel loomed brown Pahvant Butte, awaiting the return of Black Hawk, leader of the local Indian resistance in what became known as the Black Hawk War. Now I came to Old Cove Fort, a basalt keep erected by the Hinkley family as a protective way station in 1867. The church now owns it; the volunteer who gave me a quick tour would accept no donation. He led me past comfortable restored rooms opening toward a courtyard, and told me that the Indians had trusted the early Hinkleys. Never was a shot fired in anger from the gun slots in the thick lava-and-mortar walls. Though I did not know who "President Hinkley" was, or that a present Hinkley was "the prophet," my guide remained genial. I expected to be given a pitch or at least a tract when I left, but all I received was a smiling wave and a bid for safe travel.

The Beaver River carved a narrow cleft into the Tushar Mountains, where glorious gold foliage backed the pocked and whittled volcanics and their byproducts. As the sun fell behind the top of the canyon, I dove into the first campground. Little Cottonwood was a Forest Service facility but was run by a concession called High Country Recreation. I picked Site 11, beside a big narrow-leaved cottonwood and surrounded by red poison ivy, honey-colored oaks, juniper, piñon, and ponderosa. There were few neighbors, but the stream and the local red squirrel were voluble. The campground host, a vast woman in a Jesus sweatshirt, came to collect my eight dollars, and I apologized for making her walk over.

I used the last light to set up for a great stir-fry I'd been planning — road-scavenged onion, two types of tomatoes, green pepper, mushrooms, chicken, Idaho spud, and chili seasoning. Even without garlic, the result was a great improvement on the many freeze-dried meals I'd been subsisting on. Over cocoa, I wrote Thea by candle-lantern light. My tent, not yet used on the trip, remained in the car. I slept out under stars and juniper limbs, in utter silence but for the Beaver's gentle tussis. Next it was morning, the canyon's breath smelling of water and earth and leaves and early rot blended with sweet conifer terpenes. Jays called, and a mountain chickadee — now where it belonged, up beneath 11,000- and 12,000-foot peaks. The last monarch had pointed toward these mountains, a bright compass-arrow in flight.

Upcanyon into the sun, the stream purled silver through a tunnel of cottonwood gilt, the narrow road corkscrewed at last into aspens, the two species of *Populus* lined up like racks of Van Gogh's paintbrushes when he was doing all those sunflowers. I walked a ridge where purple finches fattened on seeds and a zephyr anglewing prepared for winter by cooking in the sunbeams, russet against the vibrant green of manzanita. I toasted, too, and traded my sweats for denim and chambray as the day heated up.

A little side valley bore the sign "ATV Crossing." The Paiute ATV trail, a dubious honor for the Paiute tribe, cut across the territory. Staring after a dirt bike's tracks in the fragile turf, I saw a sudden aura, quite a bright one, competing with the aspens. Parked, with my head back, I remembered when I had first had this sensation and brought it up with my ophthalmologist. "What does it mean, Doc, when someone unzips the sky before your face, a bunch of sparkles falls out, and you can't see around them?"

"Great description of a scintillating scotoma," he said. "You have classic migraine." Since I seldom had headaches, I was surprised, to say the least. But migraine manifests in many other ways. For me it was mostly the sparkles; I'd look at someone, and half the face would be gone, and I'd feel a sort of physical implosion for a few hours. The doctor said the good news was that with age, my hardening arteries wouldn't be able to dilate so much anymore and bum the brain. And in fact, I seldom get the auras anymore.

So why on this trip had I had three after none since a cluster in Sweden years ago? Was it the harshness of the high-elevation light, similar to that of a northern Swedish midsummer? All the late-night reading by Powdermilk's dim overhead light? Or was it the scintillant brilliance of the colors? Normally, my eyes catch every movement of blowing leaf or flitting junco. Earlier, a chipmunk in an oak had flicked its tail with each peep, and I spotted it right off. But if a monarch flew in my face just now, I'd have to look hard through the sparkles and the aspen glow.

A side track wound up to Big John Flat, where a truck or two of young Utahans were camped. Like them, I was loath to leave these fresh heights for the hot flats, and I was as hopeful of seeing monarchs up here as down below. Though it was way too late for any nectar but a few dandelions, I would not have been surprised to see them skimming low across the gap. Big John Flat, at 9,954 feet, was a mighty meadow, gopher-riddled, surrounded by blue spruces and the screams of Clark's nutcrackers.

Heading down, I stopped, got out of the car, and walked barefoot over a silky yellow carpet of leaves, then lay down on my back within a roundel of aspens and spruces. Some of the aspens behind me were finished and bare. The ones before me still had leaves but lost a lot while I lay there, many of them falling right onto me. The spruces were decorated with intercepted aspen leaves, like natural tannenbaums. When the breeze blew, the sky filled with leaves, having done their work, freeing their hold on life and going to ground at last. Soon, like the spruce needles, my beard was adorned with soft gold leaf.

Aspens reproduce in vegetative clones, so members of a clump tend to be all the same color. Swinging round, the breeze shook a very orange grove. When the blue air filled up with their leaves, it took me back to Sierra Chincua, Michoacán, when, lying ill on a carpet of fallen, frozen monarchs, I saw the sky fill with monarchs for the first time, and was healed. Same elevation, same conifer spires, only these flew on spinning stems instead of four wings with black veins. I'd known no more tranquil moment on this trip, and if for these moments alone, the journey would have been worth it.

The Beaver River Road, a "scenic backway," went gravel. Big Flat spread out on top of the Tushar Range in even vaster meadows than Big John. Thousands of juncos scattered from the road as I passed the old log ranger station, where I took on lovely cold water at a spring guarded by a defensive Douglas squirrel: the best water I'd drunk in this alkaline land by far. Hard to believe that it would end in the same salt sump as all the rest of the rain over the basin, with no chance of reaching a sea but by evaporating and traveling on the high currents with orange butterflies.

The Forest Service sign said "Wildlife Viewing Area." So I viewed, and saw, out on Big Flat . . . cows, of course! At the highest point it felt like the roof of the world — not really above timberline, but untimbered. I recognized from behind the mountains I'd seen looming above Clear Lake: Mts. Belknap, Delano, Holly, and Circleville. As the monarch flies, I was 58 miles and smack on the flight line of that last sighting, and virtually in line with the Bonneville butterfly as well. A lover of high country, I was stalling, hoping another would pop across the plateau and justify staying higher, longer. Three mountain bluebirds, like a slice of sky going home, flew up from the road. At the site of a recent fire, burned firs stood starkly against fresh snow and a stand of aspens that had escaped the mercurial blaze. Not so lucky, blackened aspens looked purposefully besmirched, though, in fact, fire ultimately promotes their growth.

One more flat, called Grindstone, then a last little pass, and I was over the Beavers. A common sulphur flew over the gap, dropped to check out a Coors Lite can, and carried on. A brilliant tiger beetle alighted on my shirt, crawled onto my finger, and allowed a hand-lens exam — a rare event. Normally, cicindelids skitter and fly so fast along the beach or road in front of you that a glimpse is all you get. Stippled, marked with cream commas on the elytra and the mandibles, it was iridescent, emerald shading into amethyst. Such an insect! It reminded me, with some relief, that the handsome, stinky carrion beetle had never reappeared in my car.

The land fell away into molten flows of oak, squawberry, and the reddest aspen yet. I wondered whether monarchs were ever drawn into these vermilion trees, thinking they'd found a communal roost? What a sight that would be. Thirty-four miles from Beaver it was back to a hot yellow stripe of rabbitbrush and cottonwood down to the verdant valley floor. A ground squirrel perched atop a piñon like a Christmas tree angel, after the famed nuts, no doubt. A scrub jay flashed, and I knew I was out of my beloved boreal and back to the Basin.

Back at Clear Lake, I had been just northeast of Sevier Lake, both of them surviving, ephemeral remnants of Ancient Lake Bonneville. That great inland sea covered much of the floor of the Great Salt Lake Desert, nearing the size of Lake Michigan about 15,000 years ago. When it dried out, it left a series of catchment pools in its place. One of these was the Sevier Basin, through which I had journeyed south. Now I entered the Sevier River Valley proper. The Sevier follows a twisty course, arising near Bryce Canyon, striking north around the Fishlake National Forest, then south past Fort Deseret and the Gunnison Massacre site to dissipate into Sevier Lake. In this valley I now traveled, the river was no torrent. For much of its length, alkali-heavy water barely drips through the bed. Where I joined it, at the village of Junction, it had a fair flow.

At 5:30, in the Sevier River canyon twenty-five miles north of Panguitch, I spotted a bench on a river curve that struck me as a good place to look. I pulled over, walked down along the stream, and scanned the vegetation. A monarch materialized. With my vision long since returned to normal, it stood out like a neon sign in deep space. Sometimes monarch wings look smeared with rust or smoothly painted with cinnabar. This one, in the slanting sun of a later October afternoon, was positively chrysolepic: clothed in scales of gold. And I could see every scale. It was sucking hard on a low rabbitbrush beside the river. I netted it and tagged it, #92 — the first one since Hell's Canyon!

It was a smallish, fresh, richly colored male. Holding him gently but firmly, I gazed into his polka-dotted black velvet, his deep tawny silk. I would have looked longer, but his flying sun was nearly gone, so I released him onto his bush, where he sat and fanned for five minutes. Then he launched toward the sun and the high rock gap that closed in at this point in the canyon. He climbed powerfully and was beginning to circle when I lost him into the sun, as if that's where they go after all.

I followed the river south to Panguitch, a fat green valley on one side, rabbitbrush and sage on the other. As is typical around Utah, fine old stone and brick houses stood here and there, in and out of the town; many were derelict now, forsaken for "manufactured homes," same as everywhere. The old places, dating from Brigham Young's peopling of the shad-scale scrub, were surrounded by domestic groves that looked like fine monarch roosts. I wondered what monarchs do when they can find no trees at sundown. Do they bivouac in sage, rabbitbrush, greasewood, or shadscale? I remembered what my friend Bill Calvert, a great monarch biologist, told me. Though he worked out their specialized thermodynamics in the winter roosts, he considered migrating monarchs to be "most flexible — total opportunists!"

Just before sunset I came to the mouth of Red Canyon, leading up into Bryce Canyon National Park. A raven, perched in the top of a ponderosa pine at the entrance, might have known the whereabouts of the local monarch roost, but declined to show me. A very weary party of Brit and German bicyclists struggled up the canyon, looking for a place to camp. I had no idea where I would be kipping myself. Signs warned me off all the likely pullouts. The NPS campgrounds and the restaurants were chock full — what must it be like in summer? I ate in a bright plastic place that offered "great Western fast food," crammed with Aussies, English, and Italians. Then I caught the last minute of a ranger talk at the visitor center, the young ranger quoting Ed Abbey in *The Journey Home:* "The earth, like the sun, like the air, belongs to everyone — and to no one." I took him at his word and tucked my little car behind some trees in a day-use parking area.

In my outlaw camp, dark and solitary, I finished reading Canadian William Lishman's memoir *Father Goose,* the basis for the film *Fly Away Home.* Lishman raised Canada goslings that imprinted on him, then led them on a migration with an ultralight aircraft, from which they returned naturally. Making himself mobile in four dimensions to my three, he managed to migrate with geese in a way I could never do with monarchs: "to experience the world as birds do," he wrote, "climbing, gliding, and diving through the ocean of surrounding air." Thinking back on the orange beauty sailing Icarus-like into the sun in Sevier Canyon that afternoon, I envied Lishman. Yet while he was leading his geese to a safe haven of his choice, I was following the lead of these wild monarchs. Just look where I was already, in the red-rock canyons of southern Utah, where I had never expected to come.

The next morning, over a Denver omelet at a Red Canyon café, I scanned my maps. Carrying on with the vector the Nevada and Utah monarchs had given would bring me to the Kaibab country, then on to the Colorado. About a month earlier, a friend of mine had watched monarchs migrating in east-central Utah. Mía Monroe, a National Park Service ranger at Muir Woods, who chaired the board of California's Monarch Program, knew what she was seeing when she floated through Desolation Canyon on the Green River in the company of monarchs. Mía's field notes, which she shared with me later, showed an animal distinctly southing, along a major river course, well west of the Continental Divide. September 18, a sunny day notable for regular rapids and no tamarisk: "6 monarchs with us 3 hours, mid-day, feeding on asters." September 19, a day that began cold with sourdough pancakes: "10 monarchs in a 'flock' for two hours." September 20, pumpkin pie to start, later an unscheduled swim when her boat flipped in a hole in Gray Canyon: "Monarchs at every stop, nectaring." September 21, when somehow a pizza delivery greeted the party's arrival at the end-of-day beach camp: "15 monarchs move down river with us for 3.5 hours, many nectaring at lunch."

If Mía's monarchs stuck with the Green River, in a couple more days they would have merged with the Colorado in Canyonlands National Park, due east of my camp in Beaver Canyon. Then, if they carried on down the Colorado, they'd have reached the Grand Canyon. My southeasterly fliers, sticking the course, should strike the Colorado along roughly the same stretch. Where they'd all go next was anybody's guess.

Now flying cottonwood leaves blew across the road, almost as bad as viceroys for false alarms. I passed through Orderville, a great name for a Mormon town full of neat yards and tidy gardens. A big marigold bed held just a skipper or two. At the edge of town, barberpole canyons and the jumble of juniper hills reinstated sweet disorder. My eyes darted from sandstone wonders to roadside rabbitbrush to cottonwood color, pausing occasionally on the road.

The turnoff to the Coral Pink Sand Dunes got the motor homes and tour buses off my tail. I was able to slow to see a flaming scarlet paintbrush, then to investigate what I took for either a toad or a road-killed monarch lying on the stripe in the road. It was a desert horned lizard, the handsomest horny toad I'd ever laid eyes on and very much alive. Robert C. Stebbins's Peterson Field Guide said that different populations may be adapted to the color of their habitat, and this one's ornate lateral fringe scales were indeed coral pink. On its home dunes it would be cryptic, but not in the middle of the road, where it sat stolidly on the asphalt. I moved it to safety.

Nowhere else have the Navajo sandstones produced such dunes. The painted-lady sand was very fine, a treat for the feet, hot on top, cool an inch down. It stretched away beneath green-splotched maroon rocks — organic, alive, smooth, sensual. The White Sand Dunes had struck me as feminine, but these looked positively female, unclothed, and with a sunburn. Lovely — and scarred. As at the White Sand Dunes, the cretins who managed this place welcomed dune buggies. The BLM even hosted races here on the Fourth of July.

"Motorized Fun" read the interpretive sign, leading a person who had probably never vandalized anything to scratch it all up. Another said "Protect the Area's Scenic & Natural Landscape . . . keep to existing roads and trails." Roads and trails crisscrossed practically every square foot. I passed two three-wheelers on my way out, thrashing across a small meadow. "Keep off the meadows," said the sign. I realized that, given my day as god, I would be a vengeful, Old Testament deity, smiting right and left.

Before leaving the Coral Pink Sand Dunes, I paused in the shade of a cedar, which *Juniperus* is called locally, as in Cedar Breaks and Cedar City. I wanted to gather a jam jar of the rosy powder underfoot. My father, during travels in his later years, collected jars of sand from all over, Loch Lomond to Las Vegas. I intended to share his collection with a school someday and thought this pink specimen would make a fine addition. Stooping to scoop it up, I pricked my bare arm on a yucca spike and suddenly knew I had passed onto the Colorado Plateau. Yucca, characteristic of the Southwest deserts, is largely absent from the Great Basin. That morning, when I crossed the Long Valley Summit and left the Sevier River, I had put that big sink behind me, and entered a realm of red rock, slick walls, and cactus.

ELEVEN

Apache Gold

"ENTERING KANAB CITY CULINARY WATERSHED — DO NOT POLLUTE!" In all the rhetoric of the new watershed-based, bioregional, reinhabitation movement, I had never yet encountered the term "culinary watershed," and its meaning was obscure. Just before leaving abstemious Utah, maybe it marked the divide across which good beer and coffee would again be available. Or the onset of Tex-Mex food? I saw no evidence of either. Instead, another sign announced "McDonald's just ahead." That was no watershed.

In the distance on all sides stretched red and white and coral-pink cliffs, mesas, and monuments, swirled like whipped ice cream. Stopping in Kanab for a frozen lemonade, I noticed two large, dark, white-tailed squirrels mounted as curios in a tourist shop, another in the Forest Service office. The next town was Fredonia, Arizona, whose promo brochure featured the Grand Canyon, bison, snowmobiles, ORVs, and "the home of the White Tail Squirrel." The rodent in question is the Kaibab Plateau subspecies of the Abert's, or tassel-eared, squirrel *(Sciurus aberti kaibabensis).* A couple of months before, I had encountered the extremely handsome, nearly all-black, tuft-eared Abert's, in Colorado. Now, hopeful of spotting its white-tailed relative alive, I wound my way up onto the Kaibab Plateau.

I explored several roads in the pines around Jacob Lake, scanning for a black-and-white flash crossing my path (which is how most are seen and many, regrettably, are killed). I pulled into a quiet grove and watched for the creature as I napped lightly beneath the pines. Then I searched a wooded campground, whose well-fed host said, "Oh, those guys — yeah, they're always around. They scamper up and down the pine trees, especially over there in the D Loop. It's closed for winter." But the only one I saw was mounted on the dashboard of the host's motor home.

Coming off the Kaibab, Powdermilk and I rolled down toward the base of the Vermilion Cliffs. Their monumental face is the size and color of a major cityscape lit up by a carmine sunset. Rising from the desert floor hundreds of feet high and thousands long, the cliffs announce a landscape where the only understated feature is subtlety itself. Two months later, five young captive-bred California condors would be released here — the first to fly free in years. But they weren't here yet, and I was after smaller soarers.

When I came to Marble Canyon, the Colorado River was in shade. I saw no monarchs from the high Navajo Bridge, but upriver at Lee's Ferry, I thought for a moment that I saw late monarchs over the water. Then I saw they were *bats,* shimmering in the light reflected from the high cliffs that still caught the sun. Gary Nabhan had made the opposite mistake, so I didn't feel too bad. Running the river below a few weeks earlier, he had seen monarchs regularly, flying right down the canyon. He had been surprised to see what he thought were bats flying before dusk, until his guide pointed out that they were monarchs. "We saw ten to twenty at once, streaming down the river," he told me in a phone call from the road. "In four days we saw five hundred to six hundred or more monarchs, all the way from the Little Colorado to Hermit Trail."

After cooking dinner at Paria Beach, I bivouacked for the night at Lee's Ferry, tucked between big vans in the parking lot reserved for boaters, as the campground was way up off the river and expensive for my flagging budget. Before sleep I thought of Mía floating with monarchs on the Green River, Gary doing the same on the Colorado. I envied their lazy hours sailing along in the midst of the migrants, something I couldn't really do, traveling by tire and boot leather. Their sightings underscored my belief that the monarchs make extensive use of the western watercourses in migration. Picturing the scenes, the splash, the smell, the butterflies overhead, I made their experiences my own.

Early in the morning I awoke to a canyon wren drumming its bill on the van next to me, then singing its fluty obbligato. As I ate my oatmeal, the big striped wren explored my car inside and out — racks, hatch, seats, carpets — finding spiders and bits of food. Float trips were getting ready as the sun peeked over the rubicund rim, and customers about to float down the river for fourteen days lined up at the telephone and the john. When Mormon pioneers crossed the Colorado at Lee's Ferry, they raised fortified buildings against the Navajo attacks that never came. Today, ruins of mud-brick and red stone look only a little more organized than the tumbled rimrock lining the canyon's bottom.

The bronzed boatmen all said they had seen monarchs on the river but couldn't tell me where. The river ranger, busy checking permits and gear, couldn't tell me anything at all. I set out upstream with Marsha. Among tamarisk, willow, goldenrod, and plants I didn't know, I waded up Glen Canyon. The water was colder than I expected for a day in the eighties. A great blue heron and a redhead duck were keeping company on the far shore. A ranger putted by but did not hassle me about my net. Not long before, a collector had been convicted in federal court of poaching the federally endangered Kaibab swallowtail from Grand Canyon National Park, among other protected-butterfly offenses. They had made the Park Service skittish about butterfly nets. Since I wasn't going to be collecting, I hadn't obtained any permits, and I wasn't eager to have to explain myself.

The wreckage of a steamboat lay in the shallows, now just curved thwarts and a massive boiler dangling waterweed and algae in the current. The vessel had proved impractical, requiring more coal than it could reasonably carry, so the owners left it here to rot instead of riding it through the rapids, like the steamboat up in Idaho that shot Hell's Canyon as a swan song. A Navajo cow yodeled on the reservation side of the river; otherwise the day was silent, except for two or three speedboats of fishermen that passed, and a large motor pontoon raft that drifted down full of tourists and motored back upstream unladen, leaving a sour stink of diesel on the air. A two-person kayak slipped across its wakes.

Out of cracks in the sandstone grew common dandelion, white

sweet clover, pearly everlasting, and rosettes of Southwest succulents whose names I did not know. The asters here were tall and bushy or small and white. Giant sacred daturas unfurled silky white trumpets whose toxic alkaloids furnish divine visions if ingested in the right proportion, really bad dreams and bellyaches if not. I dodged a very long-spined cactus. Prickly-pear pads lay on the beach, collecting flotsam on their spears.

A brilliant green chrysidid wasp found the white sweet clover, as did a little, wavy-banded marine blue. Like the dainty sulphur, the blue is an immigrant butterfly in the North in certain years but has no regular back-and-forth, monarchlike flow. The ubiquitous white-crowned sparrows worked the shoreline, a mountain chickadee plucked fluffy seeds from *Baccharis* brush, and desert alligator lizards skittered over the shore. Other lizards on a coral-pink boulder were themselves coral-pink. Adapted, I wondered, or chameleonic? They did quick little pushups on the hot sandstone.

I skated along the silt. Wavy algal tresses felt like the silky-smooth hair of a Clairol ad. My feet never got used to the cold; I had to haul them out into the sun from time to time. I came to a conch-colored beach where the cliffs closed in. A shelf made a perfect little pink seat. The setting was almost surreally beautiful, placid; also a little melancholy without monarchs. Was I just *too late?* A couple walking along said they had seen monarchs recently on the Kaibab Trail and at Grand Canyon Village, restoring my hope that they might still be around.

Anglers in waders worked along behind me, so I waded and walked a couple of miles farther to a rocky point where I could see around the next far bend a little. Having been bathless for days, I had to go in. I was just dropping my shorts when a youngish English couple appeared above. "You got me," I said. "I was just about to go in."

"Were you *really*?" the fellow asked.

"Yep."

"Go for it!" he said, and they moved along, smiling. I went for it, but didn't stay in long. It was a lot colder than the John Day. The Colorado here was not pink nor red nor muddy brown — it was blue:

when the 48-degree water shoots out from beneath hundreds of feet of Lake Powell, it is glass-clear, stripped of its ruddy silt. Glen Canyon Dam stands just thirteen miles upstream from where I briefly bathed. My skin turned instantly blue, too, then as pink as the lizards when I hauled out and basked on a dry throne of rock. I was sunning dry when the Brits came back by. "It's *cold,*" I said.

Speaking to me but with eyes politely averted, they both said, "I'll bet!"

"Where are you from in England?"

"We're from Sussex."

"Pretty county," I said. "King & Barnes!"

"That's my man," he said. They laughed their way up the trail, fixed with a story of a skinny-dipping Yank with a large butterfly net who knew his Sussex ales.

Lake Powell — the very name is an affront to John Wesley Powell, the one-armed government scientist who first navigated the Colorado's wild canyons and later warned against development on the arid Colorado Plateau. "On the walls, and back many miles into the country, numbers of monument-shaped buttes are observed," wrote Major Powell in *Canyons of the Colorado* (1895). "So we have a curious *ensemble* of wonderful features — carved walls, royal arches, glens, alcove gulches, mounds, and monuments. From which of these features shall we select a name? We decide to call it Glen Canyon." Now only these 15 miles of Glen Canyon are left; the other 150 miles lie beneath the cold water and captured silt of the cruelly named reservoir.

In my mind the name evokes the gently insistent flute theme that accompanies David Brower's Sierra Club film *Glen Canyon: The Place No One Knew.* As the flute interweaves with Brower's plea, you watch the water rise in one of the most sublime canyons anywhere. Many feel that Brower's victory over the proposed Green River Dam, which would have flooded much of Dinosaur National Monument, and the loss of Glen Canyon heralded the beginnings of the modern conservation struggle as much as Rachel Carson's *Silent Spring* did. In Ed Abbey's *Monkey Wrench Gang,* it is Glen Canyon Dam the outlaw characters target for destruction.

I climbed up to a stony trail that ran tight against the base of the broken slickrock and pressed myself against great slabs blackened with desert varnish. The rock was hot against my belly. The pontoon rafts floated by again, filled with folks. When I got back to Lee's Ferry, two big yellow coaches called Harveycars (descendants of the Fred Harvey Company's old open omnibuses) waited to take the latest load of "Wilderness Adventures" motor pontooners back to Page.

The Grand Canyon begins at Paria Beach. The Park Service says this is because the Grand Canyon limestone emerges here at river level, before climbing thousands of feet far to the south as the beds dip. Or, as Ann Zwinger explains it in *Downriver,* her splendid portrait of the canyon, Lee's Ferry is Mile 0 because the Upper and Lower Colorado River Basins separate there, based on the 1922 interstate water treaty known as the Colorado River Compact. By either criterion, Paria Riffle is where the last shadow of Glen Canyon meets the upper end of the Colorado's wildest reaches. As I munched my kipper snacks and filled my water jugs, considering these matters, the river ranger pulled in behind me. "Where are you bound next?" he barked.

"Maybe Mojave, maybe the Chiricahuas," I said, and started to tell him why.

"And where tonight?" he interrupted. I told him Flagstaff, probably, though I really had no idea. He didn't care. He just wanted to be sure I wasn't going to sleep in his parking lot again. I asked when his float-trip checking season was over.

"Another month," he said. I asked what came next.

"*I* get to go boating and hiking!" he said with his teeth clenched, and stalked off.

Navajo Bridge, built in 1927–28 by Kansas City Structural Steel Company, is 834 feet across, 467 feet up. The original plaque named the governor, engineers, commissioner of Indian Affairs, but no Navajos; a new brass plaque, set in fancy stonework and rededicating the bridge for foot traffic, honors Lewis Nez (b. 1900) of the Bitter Water Clan, "a friend to everyone." He carried mail and money for merchants from Flagstaff to Kanab, employed by Cedar Ridge Trading

Post. Taking Lee's ferry on June 7, 1928, with two others and a Model T, he was lost and never found. It was the ferry's last run. Now a new bridge, designed to resemble the old one, carries cars alongside the original span. Navajo vendors and craftswomen displayed their wares at the east end of the footbridge, and I selected a "cedar berry" necklace for Thea.

Once more I walked out on the impressive high span, before the peaceful sunset this time, hoping to see monarchs on the trail of the ones Gary had watched. Then along came the yellow Harveycar. The doors opened, country music blasted the quiet, and the tourists emerged with Buds in hand into a cloud of spontaneous combustion. A few emerged from the smoke to take a quick look at the gorge before climbing back aboard.

There *were* things flying in the canyon. Half an hour after sunset, a dusky immature osprey glided in from upstream, circled a few times to gain altitude, and alighted under the new bridge. A rough-legged hawk flew up past my head. The young osprey did not look comfy, perched on the bare metal of a curved beam. Its wings were held out a little from its body, its bill was open. It looked small, lonely, and precarious.

The evening was balmy as the osprey bedded down and the bats came out. Or were they?

Leaving the canyon, I couldn't simply follow the Colorado, as there were few points of access until Hoover Dam and Las Vegas. Due south lay mountains, and then Phoenix. What seemed to make more sense was to continue southeast. Such a route would cut across the Navajo lands, south of the Painted Desert, and to the Little Colorado River, taking me toward both the Mexican border and the Rocky Mountains. If there were still any monarchs south of here heading toward California, the southeasterly route would give me a chance of intercepting them. I could always hit the Colorado again later, where it defines Arizona's western frontier, if I struck out.

I drove through the night across the Indian Nation. KYNN, Navajo Radio, played powwow music, and a Navajo girl running a gas station had hair to match the pink rock landscape. I napped around midnight along the Little Colorado, and sometime after three I camped in the Coconino National Forest on what I took for a peaceful forest road. At six, huge trucks began to rumble past at frequent intervals. One of them honked, just to make sure I was awake. Then a deafening rattle and roar resolved into the biggest bulldozer I'd ever seen. It thundered down the logging road beside my camp and onto the main haul road. God knows what they were doing up in the national forest.

Sandy-eyed, I made my way down the haul road. A tassel-eared squirrel crossed the road and ran up the piny slope. Its tail was *largely* white like that of the Kaibab, which is supposed to be restricted to its namesake plateau. While those south of the Grand Canyon generally have white borders to the tails, this one was nearly the match of all those stuffed "white-tail squirrels" I'd seen. Later I learned that intermediates had turned up. So, though I couldn't find the squirrel when I was looking for it, the squirrel came to me when I wasn't (thanks to a diesel behemoth). It is often that way in nature.

Sunset Crater was a black silhouette in the sun over a big flat of rabbitbrush. On the sunrise side, the dome was red. Indians lived in pit houses here before the crater's eruption in 1064. The flat at its base, Bonito Park, was farmed for beans during World War II. Now native blue grama grass rubbed shoulders with exotic weeds like mullein. In 1928 a movie company wanted to blow up the side of the cinder cone to create a big landslide; Flagstaff citizens objected, and in 1930 Herbert Hoover made the crater a national monument. A tail-flicking Merriam chipmunk was king of the hill on a heap of lava chunks with rock roses popping out between them. Scarlet gilia bloomed big and bright, and a plant I didn't know had pungent, branched magenta stems that glowed like little neon Christmas trees against the black cinderfields.

Sunset Crater was seductive, but I had another crater in mind to visit. Even the news on the radio that Andy Devine Days were going on at Kingman — gravel-voiced Andy was one of my favorites at Saturday matinees in the 1950s — did not deter me, nor did the Painted Desert off to the northeast. During the Great Depression, my grandfather, Robert Campbell Pyle, was laid off from the Packard Motorcar Company. Traveling west, he visited and was mightily impressed by Arizona's famed Meteor Crater. He was a man not easily impressed. I meant to see it too.

On a flat, hot, prairie desert, I passed Buffalo Range Road, Babbit Tank Wash, Two Guns Road, and Twin Arrows Road. Round blue signs announced the coming attraction: "First proven impact crater on this planet." Unlike volcanic Sunset Crater, Meteor Crater resulted from an extraterrestrial impact. More blue disks read "Prototype for study of all impact craters in our galaxy," "Training site for Apollo Astronauts," "Hear lecture on history of Meteor Crater," "See continuous movies in Museum of Astrogeology & Astronauts Hall of Fame," and "Earth's Legacy from Outer Space."

I declined the tour, lecture, and movies, and read the brochure in a cool ramada overlooking the crater. It told me that some 50,000 years ago, a huge iron-nickel meteorite hurtling at 30,000–40,000 mph "on an intercept course with Earth . . . struck the rocky plain with an explosive force greater than 20 million tons of TNT." The result was a brown pock 700 feet deep and more than 4,000 feet across, with a rim crest 150 feet above the surrounding desert. A very large hole in the ground.

What I found most compelling about the crater was not the sum of all these proportions but the *scooped* nature of the concavity. Instead of forming a straight angle of repose below the rimrock, the slopes curve organically down toward the flat bottom. Together with the marked striations all around the interior caused by rolling rocks and erosion, the curved sides give the pit the look of a spacy crater such as we see on the moon under high magnification. These traits say immediately, "This place is not solely about here."

The crater was investigated and developed as an attraction by D. M. Barringer, whose family still owns it. A "Pioneer Meteoriticist," he hoped to mine outer-space iron from the lode he supposed lay beneath the crater. Barringer put much effort and expense into drilling from 1902 until he died in 1929. By the thirties, the drilling operation was moribund. Later the Bar T Bar cattle ranch leased the place and formed Meteor Crater Enterprises, correctly surmising that the crater's commercial value lay in tourism rather than in a great metal body (which, it turned out, had disintegrated upon impact). Tickets cost twenty-five cents in the forties, ninety-five cents in 1955. It is good that GrandPop came when he did, when it was free; for two bits he might have passed it by. He surely would now, for eight dollars. Anyway, he would not have appreciated the visitor center film clip of Neil Armstrong squirreling over the surface of the moon in his expensive ATV. GrandPop's favorite saying was "Man'll never get to the moon and come back to tell about it."

I could see why my grandfather liked Meteor Crater. I did, too. My only disappointment lay in not seeing white-throated swifts rocketing over the rim of the crater, a perfect playpen for these aerial aces that I had been looking for all along, at every canyon and cliff. A swift with a monarch in its bill would have been just the ticket. Along the road out, I noticed a milkweed species growing along the verge and a pink-red mallow: monarch bread and wine. The milkweed leaves were waxy, thick, green, and leathery, almost succulent; they had the usual golden aphids, ants, ladybirds, milkweed bugs, and some feeding damage, but no monarch larvae. Yellow daisies hosted common and orange sulphurs, and the first variegated fritillaries of the trip, a new orange butterfly to confuse things. They were all having a hard time with the wind, which took the flossy milkweed seeds onto the desert air.

Back onto the Joads' Road, old Highway 66, now renumbered (and mostly subsumed by) I-40. It took me through Winslow; at the corner of 2nd and Kinsley, a sign over the porch of *The Advertiser* showed a babe in red sitting on a truck tire with the caption "Standing on the Corner IN Winslow, Arizona." From there the countryside rolled by like a strange comic strip. The Black Rabbit Trading Post offered scorpions in Lucite and cold cherry cider, but the Black Rabbit ride (a famous fixture of Route 66) was down for repairs, as it had been for decades. Rock shops proliferated like tanning parlors and espresso stands in the Northwest. In Snowflake, signs said that Jake Flake was running for state representative, and Frost Flake was mayor of "Belly-button USA: Home of Navel Land & Cattle Co. Altitude: Natural High. Population Growing." Snowflake had a huge limestone Mormon temple, and local realtors who styled themselves, humbly, Landmasters. In Taylor a sandstone-slab welcome sign read "Think Positive! Share a Smile." The next town, maybe more down to earth, was called Show Low.

After many miles of such, I needed a walk. Gaillardia, also called Indian blanket flower or firewheel, and a fine nectar plant, appeared along the roads. I took a path to Pintail Wetland, constructed through creative treatment of Show Low's sewage. It was a kingdom of coots. A generous aster, an inch and a half across, and a very blue-bellied fence lizard colored the tan, rocky ground. I heard a modest call and mothy whir and tracked down a gray vireo among the junipers. Somber but sprightly, it was a new acquaintance for me. Refreshed, I happily skipped the big casino in Show Low, content to have enjoyed the byproducts of its organic output in the Pintail Wetlands.

Apache lands began in piny hills, then dropped away into the Salt River Canyon. Greeny pink strata of lichened, crystalline rock fell away thousands of feet to the river in a stunning gorge. The Black and White Rivers combine to form the Salt, which picks up the Verde east of Phoenix, runs into the Gila at Gila Bend, and (what little is left) reaches the Colorado at Yuma. Here in the canyon, the Salt was the border between the White River and San Carlos Apache Reservations.

The road switchbacked down, crossed, then climbed the other side of the canyon. I decided to spend the night at the bottom, despite the screech of air brakes and the stench of burning brake linings from the near-constant caravan of semis crawling in and out of the abyss. Between trucks the night smelled sweet, like petunias. In fact it smelled, and felt, almost tropical. Something about the place seemed right, this delirious hatchet-cut of a river that crossed half of Arizona, a clear corridor from the mountains to the deserts.

The alternately acrid and fragrant air was still. A fellow drove up and asked, "You broke down, Bud?" I told him I was just cooking dinner. He leaned against his hood and smoked. He seemed to want conversation. I didn't, but I was working up to it when a raccoon slinked by and, in the beam of his headlights, crawled right inside the raccoon-proof trash can. The man looked alarmed, got back inside his car, and drove away.

Many moths came to my headlight as I brushed my teeth: white crambids, little owlets, and a beautiful micro-moth with black-banded orange forewings and orange hindwings, half an inch across, much like a miniature monarch — en emissary from the emperor? The night was rife with grinding gears, comings and goings, animal sounds. But I slept hard until the sun rose on a day when I would be much in the presence of monarchs. It was a day that began with palms and went on to yuccas, agaves, and hackberries as I roamed down along the shore of the Salt River.

The Salt ran muddy, and the bed was reddish, sleek, cracked mud laden with tracks of raccoon, coyote, maybe ringtail. Little falls and riffles foamed beneath high, layered, and broken walls of many hues from sienna to ocher to buff. A depth gauge on an old concrete flow tower by a cable car went up to sixteen feet, but stood now at about one. Just as at Marble Canyon, the old art deco silver bridge was now reserved for pedestrians, and its new twin, red with Apache designs, for vehicles. I walked among the river cobbles and boulders, using them as stepping stones across the flow. An Apache family — mother, father, kids, and grandma — climbed the steps down to the shore and settled in by the water. More Indian kids hooted from the bridge. A white boy came down to the shore, and his dad yelled, "Don't get your shoes wet!"

I sat on a flat slab of sandstone, watching for butterflies, as a cool aquatic breath rose off the river. And here came the first streaming butterfly! A monarch? No — it was a queen, the trip's first. Now that I had entered their range, there would be two species of milkweed butterflies to tell apart, as well as the viceroys, which are dark in the South to mimic the burnt-cinnamon queens. *Danaus gilippus berenice* bears a close relationship to the monarch and a fairly close resemblance as well, especially beneath. But they are smaller (averaging three inches in wingspan to monarchs' four) and darker in ground color, more root beer than orange juice. The queens also have a different pattern of white spots, far less black scaling on the veins above, and shorter wings. Queens, too, feed on milkweed as larvae, migrate (though far less regularly or dramatically), and spend the winter south of the freeze line. The two species mingle mostly during the autumn and spring migrations.

The orange rhyolite of the riverbed was polished like a worked stone tool of the gods. From a swirling eddy I retrieved four Bud cans. With one foot raised to smash them, I looked up at a shadow, slipped, and fell. I heard glass on rock, felt bone strike stone. I unfolded slowly, upside down in the slick pool. Binoculars OK. Back OK. Not sure about my butt. Searched for glasses until I found them on my head. Marsha lay face down in the water and suffered the further indignity of carrying the Bud cans. The shadow reappeared. Slowly I stood, with both feet planted firmly. Then I followed it and found two large, dark butterflies buffeting one another by the bridge.

The dueling shades appeared and disappeared in the sun, and I could not tell if they were pipevine swallowtails or their mimic, red-spotted purples. Unpalatable, like the monarch, the pipevine swallowtail has gathered around it a whole guild of black-and-blue mimics throughout the East and South, including the red-spotted purple, the female Diana fritillary, and a plurality of the female eastern swallowtails. The red-spotted purple is a close relative of the viceroy, a member of the admiral genus, *Limenitis*, which has produced several close mimics of dissimilar, unrelated butterflies. Now one of the pair darted fully into the sunshine, and *Battus philenor* it was: the pipevine swallowtail. Matte-black, overlaid with Mylar-blue and persimmon spots, the pipevine flashed as it nectared on a ruffled pink verbena. Dutchman's pipe plants, which the dramatic red-and-black, horned caterpillars consume, trailed over the riverbanks.

And then there *was* a monarch. No mimic, no queen, no shadow, an honest-to-god monarch — my first in all Arizona. Eager but relaxed, it nectared on the sunny rim between the two bridge abutments on the far shore. I watched for a while, then ran up the steps to the bridge, crossed it, and gazed right down on the wanderer. It was a male, badly damaged, visiting one rabbitbrush after another in a clump of three. I scrambled down a gravelly slope, jettisoned the beer cans from my net, and gently caught the tattered insect.

I regretted adding to its travails — it was missing at least half of the left forewing and left hindwing, and the right hindwing was torn from a probable bird attack (white-throated swift?). It was difficult to imagine what all he had been through, how far he had come, where he might manage to get before the sum of his injuries put an end to the journey. I placed the tag on the right forewing for a change, while the subject played possum. Released on a rabbitbrush, it flew surprisingly strongly, if slowly, with the light breeze, downriver and around the opposite abutment for maybe seventy-five yards. The direction was west, but the river curved southward shortly.

Looking around for more monarchs, I prepared to net one on white seed floss, then saw that it was a rich copper queen. Whereas monarchs are larger in general and the males often exceed the females in wingspan, in queens the females are bigger. But this was a male, smaller than all but the slightest monarchs. The androconial patches on its hind wings were silvery instead of black. Monarchs do not use them very much, but queens do. The male loads his hair pencils from the alar pockets, pursues and overtakes the female, and anoints her. She slows, then alights, and he continues to brush her with fragrance molecules, finally mating with her as she grows more and more compliant.

In the Everglades I once watched male queens attracted to dodder vines, parasitic plants that look like tangled orange fishing line. Dodder emits secondary substances (pyrrolizidine alkaloids) that are highly attractive to certain butterflies. For danaiines, these are thought to be necessary precursors for the manufacture of male pheromones — without the p.a.'s, as biologists refer to them, the butterfly can't produce his full complement of sex attractants. The dodder's alkaloids, similar to those in blister beetles, also seemed to induce mating behavior between the males imbibing them. At least those particular male queens (and this is no pun) were engaged in impressive homoerotic behavior that early spring day in Florida.

Beneath the bridge shimmied a school of bullheads with barbels so long they looked like butterfly antennae. A hackberry butterfly sprang up, which struck me as odd — I knew it as a midsummer butterfly with only one generation up north — until I remembered that it was multibrooded here, like many butterflies in the South. The longer season of warmth enables the life cycle to be completed several times in a year. Snout butterflies, also hackberry feeders, shot about in acacias. Their exaggerated palpi stick straight out, producing a schnoz-like visage.

Sore from my fall, I settled into a smooth, scooped channel of the river. The water was almost warm, easy to sink into with just a little shiver, but my tailbone was tender. I lay back and let the current take me a little way, then repeated the process on my belly, sliding like an otter. On my back again, I watched queens, sleepy oranges, and blue swallowtails float over me. Gazing up at the vertical walls and the vegetation at their foot, I wondered, might monarchs cluster in the skirts of the palms? In the Prince of Wales plumed agaves? The little prickly holly oaks? The thorny acacias with tiny leaves?

Time came to dry off, pack up, and wend my way out of the gorge. Many butterflies were crossing the highway. Most of them were snouts, flying WNW by the thousands, the first mass movement of this butterfly I have ever witnessed, although it is famous for migrating in the millions. Among them I saw two monarchs, three minutes apart, crossing a mountain gap above the highway, flying NNE; then another, heading WNW with the snouts. Butterfly gardener Barbara Deutsch once suggested to me that individual monarchs might get caught up in big flights of other species. This could help to account for their appearance far to the north in major painted-lady years.

Several more monarchs flashed by, E, SE, and S, as I approached Globe, where I stopped at a Safeway to reinforce my food box. Spotting a Gossamer Bay wine display full of the promo bottleneck monarchs, I liberated two of them — one to replace the one David Branch had given me, flown away in a Utah wind, and one to give it company. I expected them to be my last monarchs of the day, but then I saw another one heading east, as both of us crossed the San Carlos River.

I picked a petal of the compass rose described by the confusion of recent headings and struck once more southeast. San Carlos Reservation signs said "Watch for Animals, Next 10 miles," "NO ATVs!" and "Bury the Booze — Not Your Friends." The first ocotillo and saguaro appeared as the route entered white, shaly mesquite hills and crossed the Gila River, lined with old-growth tamarisk. I bought Apache gasoline in Bylas and passed through Geronimo, one ancient boarded-up store and a stucco motor court. Heat waves rose from cotton fields. Fort Thomas. Black Rock Wash. Eden. I could tell when I was off the reservation when signs for the "Apache Gold Casino" and the "Peyote Way Church of God" were replaced by the Mormons' pointy spires.

Great bales of cotton rather than juniper berries lay beside the Glenbar Gin Company at Pima. The name Pima brought to mind Laurence Ilsley Hewes's evocative "Butterflies: Try and Get Them" in the May 1936 *National Geographic.* One photo is captioned: "A collector makes a sideswipe at an elusive orange-tip. Perched on a desert flower, near a tall cactus plant, is a specimen of the scarce *Anthocharis pima,* a resident of certain remote sections of Arizona. After twice searching in vain for it, Mr. Hewes discovered that the handsome little butterfly lives a life of ease, not appearing until 10 A.M., and flying for only a few days in spring." I had seen it a few springs before, below the Mogollon Rim. Lacking Marsha, I made a lucky catch in my hands, took a close look, and released it among the saguaros.

Afternoon found me on an unlovely Arizona highway with a penitentiary and a Wal-Mart backing onto the Pinalenos — home of that beleaguered rodent, the Mount Graham red squirrel. The University of Arizona and the Vatican have built a large observatory atop Mount Graham, damaging the squirrel's relictual boreal fir forest. *Boreal* sounded good in the heat, but I didn't really want to witness the infamous arrogance of the astronomical artillery lined up against the northern forest remnant and its rare endemic. Besides, I'd already been on one wild squirrel chase, and my own snipe hunt called.

After five, the air still 88 degrees, the sun under western clouds, a monarch crossed the highway heading west across flat mesquite country; then another, just north of I-10. They were flying eight to ten feet above the ground, perhaps seeking roosts near the base of the mountains. In 1960, Urquhart considered Arizona largely devoid of monarchs. Now we know of small winter roosts in canyons near Tucson. Today's had flown every which way. I didn't know what they were up to, but Arizona definitely had its monarchs.

TWELVE

Guadalupe

Cochise county is a place of ghost towns, old mines, Indian dwellings, big cowboy hats, and loud trucks. Willcox, beneath Cochise's Stronghold in the Dragoon Mountains, looked like nothing Cochise would fight for these days. But it had a pleasant green in the center, where I sat beneath an enormous elm in the warm evening. Behind me stood a bronze statue of Rex Allen, "last of the silver screen cowboys," with Koko the Wonder Horse buried beside it. A hard-living man with long blond hair and a woman in pink pants and a yellow hard hat nodded as they walked past.

I set out after dark for the Chiricahua Mountains. The longest, rudest washboard road yet, especially rough on my bruised backside, took me into the foothills. The three dancing Esmeraldas rescued, early in the trip, from the mud bank of the Sanpoil River on the Colville Reservation had been riding up front ever since. Now they shimmied across the dashboard as Powdermilk juddered over the bumps. Chiricahua National Monument was too costly with admission and campsite fees, and I couldn't swing a net there, so I bounced miles farther into Coronado National Forest. There I set up camp in Pinery Canyon, surrounded by an oratorio of insect song, the flutter of thousands of moths, and a weird, sharp animal call unfamiliar to my ears.

When I next opened my eyes it was to dusty bright sunshine falling on pines, oaks, and tall white junipers. A pair of Arizona sisters, chocolate-and-white-banded admiral relatives with mandarin-orange wingtips, patrolled the wash before my camp. The female was looking to lay her eggs on the oaks, the male was looking for her. A woolly, pied robber fly big enough to be their match stood by.

I'd come to the Chiricahuas partly because they lie smack on the southeasterly transect I had been following all along and partly because they are a good place to see several species of fall-flying butterflies other than monarchs. Foremost among these was the Terloot's white, or Chiricahua pine white. It is closely related to the pine white (*Neophasia menapia*), a common butterfly of temperate coniferous forests. Considered to be rather primitive butterflies, the pine whites have a weak and flimsy flight. The white-lined green larvae feed on the needles of pines and Douglas-firs. While not truly migrants, pine whites build up in enormous numbers in certain years, filling the skies, waters, flowers, and spiderwebs with black-veined white wings. Other than winter-massed monarchs, I have never seen butterflies more abundant than an outbreak of pine whites in Idaho's Payette River canyon. The much rarer Chiricahua pine white is restricted to Arizona and Mexico, Old and New. I first learned of it in that same 1936 *National Geographic* article that mentioned the Pima orange-tip. "When you motor through Tombstone," wrote L. I. Hewes, "you will also see at the high school some of the work of the late Arizona collector, Biederman. There are his precious specimens of the very rare *Neophasia terlootii,* a black and white southern cousin of the more northern Pine White. The female of *N. terlootii,* however, is brick red!"

I'd passed up Tombstone, though, Biederman's butterflies being long gone. Still, I very much wanted to see *N. terlootii,* particularly since that brick-red female has often been proposed as a possible mimic of the monarch. At the mouth of Pinery Canyon, I started up Pinery Trail #336 toward Horsfall Canyon and Iron Spring, under pines, oaks, and sycamores. In the dry creekbed I spotted a white and asked, What's a cabbage doing up here? Then its weak flight jogged me, and I realized it was a male Terloot's, or Chiricahua, white. I managed to catch it and was about to remove it from the net when a female appeared too. I netted her as well, and was able to tweeze and examine both sexes together. The black cell, the substantial red on the outer ventral veins, the lovely deep orange of the female . . . *Neophasia terlootii,* all right.

Released, the male flew floppily upcanyon; the female landed in a pine tuft backed by brilliant red creeper. When she took flight, it was up into a bright yellow-orange sycamore, as if she were aware of her color and had picked the most becoming setting. Minutes later, hiking up the wash, I thought I saw a viceroy, but when I netted the butterfly I found that it was a bigger, fresher female Chiricahua white. She was as bright as fresh-squeezed orange juice, with rose and cat-black on her wings, scarlet near her body — and *big* — nearly three inches across.

Since I took her for a viceroy, a known monarch mimic, perhaps the female *N. terlootii* really does impersonate monarchs. It still seemed somewhat unlikely, since I didn't expect *Danaus* to occur in *Neophasia*'s piny haunts. For mimicry to evolve and to continue working, one or both members of the pair need to be distasteful, and the model and the mimic must overlap substantially in range — or at least encounter the same predators if migration is involved. Migratory birds educated elsewhere about monarchs might be fooled by a mimic here. Perhaps these fiery whites warn off birds on their own account; to my knowledge, no one has ever tested them for unpalatability.

At any rate, they made a break from the ubiquitous cabbage whites. I recalled how impossible this butterfly seemed when, as a boy, I viewed her in W. J. Holland's standard, *The Butterfly Book.* Now here she came again, sailing through the pines of Pinery Canyon. Walking back down the wash, I passed a patch of yellow composite flowers in the sun, with several other pierids visiting them. Sleepy oranges nectared on a frilly little purple phlox with a long, gilia-like tube and blue spotty nectar guides and stamens. Then I swept the daisies and found male and female tailed oranges and a nacreous, heliotrope-and-silver-freckled giant cloudless sulphur, all at once in my net.

No sooner had I turned out the bright flutter of sulphurs than a red-bordered brown came alongside me. This was the second of four butterflies I'd especially hoped to see here. The browns, also known as satyrs or ringlets, are among my favorite butterflies, with their softly striated patterns resembling fine pelage. *Gyrocheila patrobas* is the only U.S. species of the tribe of pronopheline satyrs, a group widely radiated in the Andes. The female I had in hand was velvety brown, with a question mark of white eyespots on the forewing, the hindwing edge fox-red on top, ponderosa-purple beneath. On my finger, she sucked my perspiration, as had the female white. Placed on a daisy, she immediately nectared.

Little brown satyrs flickered up and down the dry streambed. I caught one and carefully placed it in a protective envelope to compare with field guide illustrations. Then a brighter red-bordered brown arose, after invisibly basking beside a polychrome chip of sycamore bark. He too went for the sweat. I said, Here, you want sweat, try this — and placed him on my damp forehead. The soft tickle on my skin, like the butterfly kisses between my fingers, was very agreeable. He settled on my sweatband and dug in. When I got to Powdermilk, I saw my reflection in the window: red-and-white flowered dewrag over greasy hair, gold terry-cloth sweatband beneath, like a burnoose crowned with a butterfly as a forehead jewel. The King of Araby — and I felt it!

I took the smaller satyr out of its envelope and found it in the field guide: *Cyllopsis pyracmon,* the third species on my list. Flying, it looks like nothing more than a scrap. Up close, it is reddish fawn patched with silver, cloudy orchid, and deep blue in a field of iridescent scales that Nabokov called *pervulent.* The Audubon field guide reminded me that I gave all three of these butterflies the English names they now bear: Nabokov's satyr, red-bordered brown, Chiricahua white, all life butterflies for me.

As if they weren't enough, a monarch came flapping through the Pinery. A *real* monarch — no female *N. terlootii,* no viceroy, no queen. It was flying purposefully at twice eye height, aiming ESE, 150–160 degrees, on still air. I watched it well for a hundred feet or so before it rose over the pines and sycamores. Then, right on the heels of the biggest migrant butterfly, came one of the smallest. Momentarily I detained the bright female dainty sulphur — olive, yellow-orange, lemon, black. A monarch tag would have covered an entire forewing.

So here was a monarch heading southeast again, at midday. Following its path would take me to Portal, Arizona — where a monarch tagged in Goleta, California, turned up in April of 1988. No one butterfly makes a trend, but that recovery causes one to ask whether Arizona receives spring monarchs from both Mexico and California. For now, this Pinery monarch suggested both the need to keep an open mind on mimicry, for monarchs and Chiricahua whites obviously *can* overlap in habitat, and a way for me to go.

Clouds came full over, and thunder. The steamy, sweaty morning gave way to a nicely cool afternoon in the midseventies. Big, shy jays glided over the Pinery Loop Road now and then but resisted approach, and American robins migrated through mistletoe-hung forest. A dogface sulphur, its black border indented to resemble a poodle's profile, and a dainty sulphur nectared together in a composite border, two of the half-dozen sulphurs that expand into the north with the monarchs in the spring but do not return. Some, obviously, stay behind, and these were their offspring. At 7,600 feet, the crest of the range, the rain let loose, the first I'd seen in many weeks. "Use Extreme Caution" read a warning sign: "Recent Forest Fires Have Increased Flash Floods on All Roads." But the rain was short-lived. I was too late for the Southwest monsoon.

Two little gray deer bounded across the road. I willfully mistook orange-needled evergreens for monarch trees like those in the mountains of Michoacán, but these trees were merely diseased. A vista opened onto sharp-ridged, sharp-shadowed ranges off to the northeast, a big tusk of a peak shining in a stray sunbeam, and the Peloncillos off in a distant purple dust. Cochise Head loomed off to the left. Now *this* country looked worth fighting for.

The mistletoe on junipers looked like dwarf mistletoe, which feeds the larvae of small brown hairstreak butterflies, while that on the oaks looked like real Christmas mistletoe, which would harbor the great purple hairstreak. So many unfamiliar trees — alligator-barked junipers, bays, long- and short-needled pines. When I had packed the bulging book box on the back seat, I had forgotten to put in a field guide to trees and had regretted it often. I met the coy jays again, which, along with a smaller, darker Steller's jay or two, were mobbing a raptor. Thus occupied, they allowed me a look at them, and they turned out to be gray-breasted or Mexican jays *(Aphelocoma ultramarina).* I could see that the specific epithet was well deserved, as the wind parted one jay's lush gray cravat and it flashed ultramarine wings, contrasting with the Stellers' deep azure.

East Turkey Creek ran down to a junction: Paradise, three miles left, or Portal, nine to the right. I turned right, a route that also offered Sunny Flat and Herb Martyr, and came to the Southwestern Research Station of the American Museum of Natural History, tucked among dramatic orange and lime rhyolite ramparts above Portal. None of the scientists were around, but the office manager, Shirley Cox, had noticed monarchs moving through. She'd seen several passing the station southeasterly in the past week or two, migrating to Mexico, she presumed: the border was only sixty-eight miles away.

An Arizona sister drifted by languidly, and a pair of gaudy acorn woodpeckers stuffed oak-nuts into a pine trunk. Down along Cave Creek, white sycamores and blue junipers shone against ham-pink pinnacles. The tuff walls were riddled with Swiss-cheese caves and nooks, custom-built for the renegade Chiricahua Apaches, who hid out here with Cochise until the army finally flushed them out at great expense. I hiked up to gain a view of Cathedral Rock, a remarkably impious pillar for such a name.

At the ivy-clad, white-striped, pillared ranger station for the Coronado National Forest, I learned that the coppery-tailed trogons were long past and the elegant trogons were not here yet, so the birding was quiet in this premier birdwatching location. Among a display of live rattlers of various species, along with kings, gophers, and coral snakes, lived a handsome big black-and-orange-beaded Gila monster from three miles away. I took a good look, in case I never managed to see a wild one. Few other toxic animals share the monarch's color scheme to such advantage.

The road out of the forest teemed with wildlife too, alive and dead, venomous and not. A gigantic red and green grasshopper told me I was truly nearing the tropics. Then, by Cave Creek Ranch, a small tarantula was crossing the road. Much more legs than body, it made stately progress, but it really scooted when I touched its leg. I saw it safely across. A well-mashed road kill was even more exciting. Closely examined, it proved to be *Agathymus evansi,* one of what Nabokov's Ada describes as "the most noble animals in America, the Giant Skippers" — butterflies that fly at freeway speeds and burrow as larvae into the rootstocks of agaves and yuccas. A good bottle of tequila or mescal is supposed to contain one of these caterpillars, though these days the distillers usually substitute beetle grubs. I had never before seen a giant skipper in the wild, dead or alive. When bestowing a common name on this species for my field guide, I compounded its epaulette-like markings with the rank of its namesake, skipper specialist Brigadier W. H. Evans, and called it the brigadier. That name was later replaced by Huachuca giant skipper, for the mountain range to the west where it was discovered.

Vladimir Nabokov visited Portal in 1953 to investigate Nabokov's satyr, though it wasn't called that then. Traveling in the West each summer with his wife, Véra, who drove, VN was in the habit of collecting butterflies on clement days and working on his novels in bad weather. That summer he was writing *Lolita,* and at Portal he made good progress after rain replaced an initial spell of sunny weather. I asked a woman with a face colored and weathered like the local stone if she knew of any lodges or cottages that had been there in the fifties, and she replied that there were some near Cathedral Rock Lodge. An old-timer knew just which cottage the Nabokovs stayed in while he worked on the book. Though the locals took note of Véra playing the piano while Vladimir muttered about the weather, they had no idea of the history he was making in both literature and Lepidoptera.

I crossed into New Mexico in a sunset that blended clouds and the tops of the Chiricahuas. I stopped and leaned out to pick up round yellow squashes by the road, thinking they must have fallen off a truck — but they were *growing* there, hooked onto vines rooted in the verge. Big red-and-black sucking bugs clambered over them. It is good that I did not try to eat my road-find squashes. In *The Desert Smells Like Rain,* Gary Nabhan tells of biting into a *Cucurbita foetidis-*

sima: "the taste was so terrifically bitter that my tongue muscles went into shock. I spat the pulp out and ran for water." An old Papago Indian woman told him, "It *used* to taste just like an apple. Then Coyote came along and he *shit* on it." And that is how the coyote melon got to taste so bad.

Rodeo, New Mexico, was a cute little roadside burg built by Phelps Dodge in 1902 to serve the rail line linking El Paso with Bisbee's copper mines. Southern Pacific's *Golden State Limited* and *Californian* passed through here until 1952. Now there's a gallery in an old church, a grocery, a salon, and a souvenir store sporting a big Black Cat Fireworks poster: "Black Cat is the Best you can get." I doubled back on State Line Road to try to see a monarch in New Mexico. Instead I watched a black merlin slice into the sunset.

Back in Arizona, just north of a ghost town called Apache, consisting of a derelict stone-masonry hall, stood a handsome stone pillar — the Geronimo Surrender Memorial. Near here, in Skeleton Canyon, on September 6, 1886, Geronimo and Nachite and their followers surrendered to General Nelson A. Miles. According to the plaque, Lieutenant Charles B. Gatewood, with Apache scouts Krita and Martine, "risked their lives to enter the camp of the hostiles to present terms of surrender offered to them by Gen. Miles. . . . After two days Gatewood received the consent of Geronimo and Nachite to surrender . . . forever end[ing] Indian warfare in the United States."

Erected in 1934 by the city of Douglas with federal funds, the monument was about fifteen feet tall, a blunt obelisk set on a stone pedestal. It was built of sparkly cobbles and pebbles with several serving-platter-sized thunder eggs mortared in. The only sign of vandalism was broken glass in some of the little windows protecting the interior of the geodes. I put my fingers inside one to touch the crystals and, feeling web instead, whipped my hand back out. When I looked with my flashlight, there was a beauty of a black widow, red and black like Gila monsters and monarchs, her hourglass displayed, embracing her egg sac.

Two young coyotes, no doubt on the lookout for melons to shit on, didn't flinch when Powdermilk approached but trotted down the white line in my brights for quite a way. Farther on lay a coyote who had made the final sacrifice for doing something similar. I removed it to the verge. Then came a coatimundi loping along, its long slender tail slung over its back. I wouldn't have been entirely surprised to see a jaguar next, an animal that has indeed been making a reappearance north of the border in recent years.

In Douglas I took a thirty-six-dollar room in "The Last of the Grand Hotels," as the Gadsden called itself, a mile from the Mexican border. After a Mexican meal I stopped in at the hotel's Saddle & Spur Tavern, with its little blue neon sign, cowboy-pattern tables from Rex Allen's heyday, and hundreds of cattle brands on the walls. A regular bought me a beer just for requesting a Negro Modelo instead of a Bud.

The next day I woke to rain. After breakfast I had a quick look at the famed lobby of the Last of the Grand Hotels, lively with bilingual repartee. The Gadsden, named for the Gadsden Purchase, was built in 1907 — just twenty years after Geronimo, and before statehood. It burnt and was rebuilt in the 1920s. Plastic plants and all, it still had grandeur. Marble pillars flanked a flaring white marble staircase, with classical statues of old warriors that certainly were not Apache braves, on the balusters. At the top hung a brooding, arch-topped oil of Cave Creek Canyon and pioneers by Audley Dean Nichols. A young woman waved out of the back of a Conestoga like a merry flapper, but the long-rifled, buckskinned scout on horseback looked grim.

The real glory of the lobby was its magnificent Tiffany windows, depicting a Sonoran scene in an eight-panel, forty-two-foot-long glass mural. The artist had subtly but vividly captured saguaros, yuccas, cottonwood, blooming flame-tipped ocotillos, prickly pears, agaves, and other flora. The plants were glazed in vibrant greens, browns, and reds, the rocks in mauve and pearl, the bright blue skies mottled with white and browny blue clouds. I looked, but detected no small flecks of orange.

Eager to get into the real desert, I packed up . . . and locked my keys in the car. I went at the small window crack with a coat hanger, while the waitresses, desk clerk, and maids cheered me on. After three-quarters of an hour, using two coat hangers grafted together with duct tape, I was able to hook the key ring and unlock the door. The sky was beginning to clear, illuminating the lobby and the stained-glass desert brilliantly, as finally I strode down the marble staircase and out of the Last of the Grand Hotels at eleven.

I was hoping to follow the border west to Palominas, then explore the riparian reserve on the San Pedro River on foot until sunset before dashing north to Tucson — all in six or seven hours. Looking for the way out of Douglas, "Gateway to Mexico," I got lost on a street that dead-ended at a corrugated metal wall. It dawned on me that this was the border, just as I noticed three white and green trucks and three Border Patrol officers lolling in the sun beside them. Seeing me, they perked up and looked quite eager to make my acquaintance. I spoke first: "Gentlemen, let me state my business."

One of the men joked that his partner knew all about butterflies (*mariposo* means gay as well as butterfly in colloquial Spanish). The third said his children chased them, and they'd had plenty of monarchs. "Big orange buggers" — he gestured with spread fingers — "all around the house. But where you oughtta go is Guadalupe Canyon," he said, gesturing east over his uniformed shoulder. "That place is a natural pathway for all sorts of birds and bugs."

"Gateway to Mexico" had an entirely unintended meaning. Several months later I read in the *Washington Post* that the notorious Joaquin Guzman Loera drug organization had constructed a tunnel from a home in Agua Prieta, Mexico, under the border to a warehouse in Douglas. Tons of cocaine were smuggled through it until the Drug Enforcement Agency received a tip and shut it down. The fact that the tunnel came up quite near where we met might help to explain the officers' immoderate interest in me. I'm afraid I was a disappointment to them; even if they had searched, the worst drugs I had on me were four coyote melons full of cucurbitacins. But as it happened, that casual tip from a Border Patrol biogeographer would prove extraordinarily important to me.

So I traveled east, not west, out of Douglas, toward the San Bernardino National Wildlife Refuge and, eventually, Guadalupe Canyon, if I got that far. The dirt road, parallel to the border, bore into the humpy Sonoran desert two or three miles north of Sonora. It was full of grasshoppers — long, sleek, coal-black ones and short, fat lubbers with pinkish wing cases and blue faces, weighing an ounce or two each. They came out to feed en masse on road kills of their own kind, many of them then adding to the carnage.

The countryside was studded with big rock outcrops like bad molars. The music on the radio was all Mexican. A detached car door had become a "no trespassing" sign, and old-fashioned windmills pumped away as rain played lightly on the windshield, then grew heavier. A maybe-monarch crossed north in front of me, likely seeking shelter from what quickly became a short-lived downpour. It was probably a queen.

I was just thinking that these fluffy, white-flowered bushes (*Baccharis,* desert broom) were what Lincoln Brower and I once saw monarchs swarming on in Sonora, when I looked over and saw big orange butterflies all over one of them. I leapt out, thinking I'd finally found a bush burning with monarchs, but as I approached I could see they were queens — and hundreds of other butterflies, all on the one broom. In the sun, back out already, I counted twenty species, including queen; pipevine swallowtail; dogface, sleepy, tailed, and dainty sulphurs; gray and Sonoran hairstreaks; pygmy, solitary, ceraunus, and marine blues; pale Mormon metalmark; Empress Leilia hackberry; Virginia lady; variegated fritillary; bordered patch; snout; long-tailed and big black skippers — practically everything *but* monarchs.

Half a mile on, I slow for what I think will be a large queen, and it is a monarch for sure, crossing the road. It is flying purposefully SSE at 160 degrees in overcast and still air. While I do not actually see it cross the border, it will reach that invisible line in about ten or fifteen minutes — fifty miles west of the Continental Divide. I am thunderstruck by the thought that this one butterfly holds the power to contradict a major myth of American natural history: *that all the monarchs west of the Rockies' crest migrate to California, only those east of the Rockies go to Mexico, and the two don't mix.*

This monarch may or may not be bound for the wintering grounds in Michoacán, but it is *not* heading toward California. Nor was the one I saw yesterday on a similar trajectory, coming through the pines in the Chiricahuas. The myth may very well be true on the whole; maybe *most* western monarchs flow to and from the west coast. But not *all* of them. Some, apparently, fly to Mexico — the evidence is here, before my eyes. Another iconoclastic monarch crosses the road, goes down to nectar. Sandaled, I search carefully among the cacti and scorpions, queens and variegated fritillaries, but it is gone.

At the historic ranch of John Slaughter, famous Texas Ranger and sheriff of Cochise County, virtually on the border, an old cowboy who seemed to know a king from a queen reinforced my observation. "Yeah, they come through here," he said, and directed me to the San Bernardino Refuge. Part of an old Spanish land grant, the refuge seemed to be chiefly mesquite. I thought of walking Black Draw toward Mexico, but it was posted "No Entry." It was also raining again; I did not linger.

Guadalupe Canyon sounded intriguing, but it was twelve more miles away by muddy, rutted dirt road. Having seen two monarchs — and the best butterfly bush in my life — I decided to head back west to Palominas and, while there was still light to see it, the San Pedro. Lightning lit up the Rockies, and the thunder struck where I was. As I turned back out onto Geronimo Trail, a monarch — in profile but close and clear — crossed my bow heading south at 180 degrees. That made three. There was a rainbow over Guadalupe and a "no hay paso" sign. Impulsively I turned east after all.

Barn swallows were moving through, too, and I wondered if they could be the ones I had seen staging at Salt Lake. In *Rites of Autumn,* Dan O'Brien writes of a flock of Canada geese, "It was unlikely but not impossible that they were the same flock that had shared the pond with us in Montana." I felt the same way about the monarchs I'd followed all this way, each time I saw them: not likely the same ones . . . but not impossible.

A beautiful dark gopher snake, over three feet long, "rattled" its black tail when I touched it. While I was stopped, a sleek Jeepful of young couples paused to ask if I, and the snake, were okay. "Are you going to Guadalupe?" the driver asked. I said I might. "It's nice," he said. "Private, but it's open to walkers."

The road grew rugged and rocky for the last few miles. It climbed over a lip of land, then dropped abruptly toward the canyon. An understated sign on a substantial gate said you could enter if you were not hunting, trapping, or on an ATV. And so at five P.M. on October 14, the sky clear and the still air at eighty-nine degrees, I walked into Guadalupe Canyon after all.

I later learned that Guadalupe Canyon Ranch, owned by rancher Drum Hadley, is one of some thirty-five private holdings that have banded together with environmental groups and government agencies to improve the rangeland and to keep unfragmented their nearly one million acres of open space. The Malpai Borderlands Group calls its collective lands a "working wilderness" and its political position the "radical center." These landowners want to keep their way of life, with cattle, while maintaining and enhancing biological diversity. A brochure for the group says, "The sycamore-dominated riparian area of Guadalupe Canyon provides a migration corridor for many Mexican wildlife species." I'd heard that somewhere before.

Now I walked into a broad sand wash overhung by those sycamores, and cottonwoods. The trees were huge and cavernous; Geronimo may have known some of them. A prickly pear grew twenty feet up in the crotch of a cottonwood that was nearly ten feet through at breast height. Eroded, flesh-colored rhyolite formed the canyon walls. The dry streambed was unmarked except by javelina tracks. Little chartreuse tomatillos dangled by the bank. Only the soft vowels of quail and dove dented the quiet.

The wash turned north and broadened. This was where the great wave should finally come — monarchs coursing overhead, Apache arrowheads glinting beneath. But neither showed. Just three queens nectaring on a yellow daisy clump, holding on tight in the light breeze. They ought to be monarchs, I thought. Then, no; it is right and fine that they are what they are. Gazing out at the white cliffs of New Mexico, the red receding mountains of Mexico, both so near, I realized that it was *all* all right. I turned around, toward home. I'd come as far as I was going. From there on, it was all toward home.

THIRTEEN

Buenos Aires

A LONG DUSK DRIVE with one headlight took me past the stairstep lights of Bisbee, across the forsaken San Pedro in the night, alongside the invisible Whetstone Mountains that share my Colorado ancestors' family name, through Benson to Tucson. There I swapped monarch stories with Caroline Wilson, former chief naturalist at Organ Pipe National Monument, and with poet Alison Deming, who had recently completed a poetry sequence entitled *The Monarchs.* Besides butterfly talk, Alison's hospitality included fine Mexican food; a pool from which I could reach up and pluck fresh kumquats that greatly enhanced a cold Corona; and a spare room equipped with twin kittens.

As I dropped off, the image came to me of a motorcycle with a sidecar that had been offered by raffle back at the Gadsden. A Russian Ural, it was shiny olive, old-fashioned in design, sturdy, and mighty handsome. When I slept between the borrowed cats, I dreamt of that massive motorcycle, the color of an olive in air as crisp as gin, rumbling under me. My hands were spread on the broad, high handlebars. Thea rode beside me on the left, tucked snugly into the metal pellet of the sidecar. Our nets trailed behind. We were heading south on a small, smoothly paved road. Suddenly we crossed a high, arched bridge, and looked down to see the San Pedro River lined with lush vegetation as far as we could see in both directions. The channel carried a current of solid monarchs, flapping and gliding toward the border, which was lit up in neon in the distance. I goosed the Ural, charged into the stream, and we were flying . . .

In fact, I strongly suspected the San Pedro River, southeast of Tucson, of being a pipeline for monarchs. On this trip, nightfall and my shrinking time prevented me from having a look. But in April of 1998, chased south by snow that nearly reached the border, I returned to walk this important riparian conservation corridor from Palominas nearly to the line. Too early for returning monarchs, I saw only queens, but the shady cottonwood banks looked promising. My suspicion was later confirmed by Arizona naturalist Roseann Hanson, who saw many monarchs nectaring on button willow at St. David cienega on the San Pedro in August and, on November 3, one ratty monarch flying high and strong southeasterly upriver toward Mexico. Both sightings occurred within forty miles of the border.

West of Tucson, mockingbird and cardinal song floated through the car window. Pipevine swallowtails floated, nectared on red lobelias, and flashed the same blue-black as the perky, crested phainopeplas with their whiplash call. Two likely monarchs crossed the road; in their wake, a roadrunner paused, bobbed its tail, flashed its red neck, and *ran.* Then southward into the Altar Valley, where a big wash lined with desert broom's woolly cream blossom pulled me over. Many of the same butterflies of the day before were there, along with brilliant, quicksilver-spotted gulf fritillaries and my first Palmer's metalmarks. Several species of yellows swarmed over the pink verbenas. Datura petals, wilting in the sun, looked and felt like the white kid leather of my father's Masonic apron. A cluster of five bushes hosted hundreds, maybe thousands, of snout butterflies. Had there been any monarchs there, they would have been hard to pick out amidst the big, bright butterflies and huge orange-and-black tarantula hawk wasps.

But they had been around recently, as I was about to learn. In the small border town of Sasabe, I stopped at La Osa Bar. Just a mile from Mexico, it offered no Mexican beer, so I sipped at a Coors and asked the manager what she knew of monarchs. I was amazed when Georgia told me, with a light Latina accent, "In September, the monarchs came through in the hundreds — a trickle first, then for a couple of nights a lot." They roosted in a mesquite behind the barn, she said, then flew directly south the next day. "I heard on Mexican radio in early October that the monarchs were arriving down there," she said. Did she know queens? "Oh, yeah, they're different. They're here every year, a lot. But there were many more monarchs this year than in recent years." Georgia took me out back and showed me the monarch mesquites.

"I remember monarchs coming through some years, but not this one," said Carol, a Customs and Immigration officer at the Sasabe crossing. We talked in the shade while Virginia and painted ladies lunched at planter flowers. She told me of the birds and butterflies that came to her birdbath, how a golden eagle had stayed for three days. She too was quite sure about monarchs coming through in the fall, though she had missed the mass flight that Georgia saw. Between the two, here was more evidence of western monarchs crossing into Mexico. What I had seen near Guadalupe evidently was not a fluke.

I'd thought of dipping into the Mexican village across the border here, also called Sasabe. "Don't bother crossing over," said one of the other guards. "There's nothing there." A Mexican boy biked through the border with an ID card and a jug and returned with fresh water, a daily ritual.

Back up the Altar Valley, a western long-nosed snake with pink, yellow, and black bands a lot like those of a coral snake (a likely mimic) had failed to make it across the road. Way too many people do not brake for snakes. I turned into Buenos Aires National Wildlife Refuge. The biologists and managers knew nothing of monarchs, but they put me on the phone with Roseann Hanson, who ran a butterfly bait station in the refuge unit known as Brown Canyon. She told me she had seen one just the other day. We arranged to meet in the morning.

At dry Aguirre Lake stood an old corral left over from the erstwhile Buenos Aires Ranch — double walls of mesquite logs with branches stuffed in between. Coyote melon vines twined up the fading green loading chute, the fruits like so many chartreuse tennis balls. Big black darkling beetles lumbered down the roads, and loggerhead shrikes perched atop yucca stalks. Along Arivaca Creek, more than a hundred sulphurs of five species as well as a few blues were puddling for dissolved salts at a damp spot under a ledge. As I perused a patch of pearly everlasting, a big Virginia lady flew onto it: a butterfly and its host plant. The cause of another name for it, painted beauty, showed clearly: intense pink and mauve below, with lox-orange uppers.

Then, about five in the afternoon, I saw a *big* orange butterfly up in the cottonwood tops that would *not* give me a look. Sulphurs, a hairstreak, and hackberry butterflies were also up there, perhaps going for aphid honeydew or fixing to roost. The big orange shimmy-settled onto a branch just like a monarch, but when I got a fix finally, I saw only queens, roosting as monarchs do. Well, they seem to migrate, though to a much lesser extent than their relatives; why shouldn't they overnight communally as well?

The next morning in Tucson, I watched one of Alison's kittens pounce at a giant swallowtail nectaring on lantana, and miss. It is the biggest butterfly in North America, even greater than the monarch in wingspan. Before the day could heat up, I drove southwest again. Beneath the great stump of sacred Baboquivari Mountain, I followed directions to a mailbox forest, through hunter gates, and through the dust to a white gate. Roseann was there to meet me and led me another mile into the canyon's throat.

Roseann and Jonathan, her husband, were volunteers for the U.S. Fish and Wildlife Service at Brown Canyon, a satellite of Buenos Aires NWR. "Roseann slid us into this," said Jonathan. "She said, If you ever need anybody to live up there, let us know. They said, Can you move in in May?" That was eighteen months earlier. In the cool shade of their keepers' house, Roseann gave me fresh lemonade flavored with prickly pear. They described Brown Canyon as a wonderfully diverse enclave, with rarities like zone-tailed hawks, thirteen species of hummers, buff-tailed nightjars, and the fifth largest rhyolite bridge in Arizona. Desert scrub blends to piñon-oak in thirty miles, and a side canyon has four habitat types — grassland, piñon, saguaro, and mesquite/sycamore/pine.

Few butterflies were visiting the rotting fruit on her porch-side bait station, but she assured me that Arizona sisters, hackberry emperors, and others came daily. That spring had been very hot and dry, but on July 27, after heavy rains brought five species of milkweed into bloom, fresh monarchs appeared nectaring on them. She saw several each day through August and September. The latest monarch recorded in Brown Canyon was on October 10, cruising the porch, with no strong directionality.

Roseann led me up into the meadow. We saw queens, and then — yes! a monarch. I chased it for fifty yards and caught it. It was an old male, worn, missing the tip of the right forewing. I tagged the left, #81994. He played possum in my hand, then flew into a patch of stinkgrass, up to a little acacia, and finally disappeared into the arms of a massive sycamore. Two days later Roseann would find him still around. He might well have been of the earlier, nonmigratory generation.

We continued up a well-used mountain-lion trail to a generous stand of desert broom, one of Roseann's favorite shrubs, although many people consider it a coarse weed and root it out. Recalling the thousands of butterflies I had seen on it in just two days, I found this astonishing. Roseann asked if I'd seen a lot of snout butterflies. I told her there had been many on the trip, though my previous total had been three. "Well, this," she said as we approached the broom, "will snout you out." On one bush alone there were hundreds of the beaked butterflies, poking hungrily into the creamy inflorescences.

We hiked on up to a pond in the canyon, where Roseann had seen a puma drink. Poolside flowers attracted great purple hairstreaks, the *Atlides halesus* I suspected had been around the mistletoes in the Chiricahua oaks. Not really purple but shimmery blue, *A. halesus* is black below, with brilliant fire-engine patches on the abdomen and on the wings near the body, advertising its unpalatability to red-sensitive birds. After watching it nectar above the quiet pool in a stone basin, we returned to the cabin for another glass of cactus-pink lemonade. Over quesadillas, the Hansons told me of their hopes that the government will respect Brown Canyon's special qualities instead of developing it as a money-making super-birding site with too much use for such a compressed, narrow canyon.

On the way out, I searched steadfastly for the lunker of a Gila monster that was sometimes seen near the white gate. But as the Hansons have written of *Heloderma suspectum*, "Gila monsters spend about 99 percent of their lives underground, either resting, hibernating, or hunting for the dens of small mammals from which to steal young."

Late that night I stood in the desert of the Tohono O'odham Indian Reservation, helmeted by star-flecked black, not a single bright electric light in sight. This was no place I had ever imagined coming. In the afternoon I had driven to Gary Nabhan's house, on the western edge of Saguaro National Monument. The great guild of saguaro cavity-nesting birds was much in evidence — Gila woodpecker, gilded flicker, and cactus wren all lively and noisy. A chocolaty adult Harris's hawk watched me from a crotch of a cactus, while its two young perched in a shady palo verde by the house. Gary told me he used to avoid them, so as not to disturb them. "But now," he said, "they follow me when I run." A little later, the writer Peter Steinhart arrived, along with two research assistants, graduate students from Arizona State. In the dusk the five of us drove a long way west, to the village of Big Field, deep in the Tohono O'odham Reservation.

After the overheated day, it was pleasantly cool in the village feast house, with its open courtyard, stuccoed adobe main hall, dirt floor, long tables, and cooking pits. A Papago elder named Delores joined us. Gary hoped to convince him to accompany us the next morning on a visit to an isolated village with one remaining occupant. I asked Delores about monarchs, showing him one of my wine-bottle butterflies so he could be sure what I was talking about. He had seen them at Big Field, but never a roost, and not lately. "I used to see a lot in Green Valley," Gary translated from the Papago, "on the Santa Cruz River." He'd had lots of water then and grew barley, cotton, lemon grass, and pecans. The word he used for the monarch was *ho' ok mal,* which Gary said applies to butterflies in general. "That's a nice butterfly," he said in English, fingering the cardboard monarch. "I like that butterfly." I gave it to him, and he seemed pleased.

After finishing the bottle that the monarch came on, we tossed our bedrolls in the courtyard surrounded by a living ocotillo fence. Indian dogs and great horned owls woofed and the stars rang over the broad bare yard between the feast house and the old adobe church. Gary joined me in the dark, watching bats. On his way to his house, Delores asked Gary what he'd done with the moon. "I thought you did it," he replied.

Gary led the morning drive west, through places called Why and Ajo, and off onto little dust-and-rock roads through saguaros and chollas and palo verdes, fording sand washes. Driving fast, he lost me once at a Y, and it was only by chance that we met up again. At last we reached Chico Suni, almost a ghost village in a no man's land next to the Barry Goldwater Range on the Cabeza Prieta National Wildlife Refuge. Its solitary resident, "Tulpo" Francisco Chico Suni, lived in a little collection of mesquite, pine plank, plywood, ocotillo, saguaro rib, and corrugated iron huts, bound up with baling wire. There was once a village of forty families here, but the other residents had moved away or died off.

A couple of old bedsteads lay under a ripple-green fiberglass ramada, and old armchair frames stood out in the sun. On a cobbled table sat a milk can, old gas cans, a pan of salvaged, rusted nails and screws, a barrel, a washtub, an old rusted sewing machine. Other little washing and cooking ramadas shaded parts of the gravel yard. An overturned wheelbarrow sheltered a pile of ocotillo kindling, in the base of an old iron cookstove, from the rare rain. Another barrow was tipped over a gravel pile with a sifting screen braced over it. Near the hut sat an adult tricycle with a basketful of cans and bottles. Old bits of iron lay here and there in the gravel, ruins of ancient cabins. Stumps of beams, rotted or burned, emerged at ground level. A yard, fenced in rusted bedsteads, springs, scrap wood, wire, and whatnot, contained a big agave for making maguey. Two children's tricycle frames poked through innumerable metal vessels.

Gary was working with the Tohono O'odham alliance Indian elders to have the village designated a tribal historic site to protect it from the big federal powers. As the students took measurements, he set about mapping the site, skittering from place to place, getting down coordinates. Meanwhile, Peter, who was working on a book about hermits, attempted talking to Tulpo with the help of Delores, who spoke a different O'odham dialect. Soft voices issued from the main hut of purpled tin. Both Indian men wore jeans, Delores a snazzy

Tohono O'odham cap and a plain blue shirt. The hermit wore a shirt that once was green and leather pull-ons as weathered as his cinnamon skin. "Ho," the men said. "Ha'o."

I wandered about and spotted a queen passing through. A blue-gray gnatcatcher buzzed the mesquite above an old covered well whose wooden yoke was furred with use, though the pulley was rusted. A cardinal crossed a sand wash, leaving a pink feather. Small caches of rocks, bottles, wire, and plastic sprouted here and there among tufty little pussy willows and creosote bush. Broken glass and chips of mica decorated the base of a six-armed saguaro, whose arms held rainbow spider orbs in their elbow crooks. A dead saguaro stood starkly nearby in the gathering heat, its dried and liberated ribs bowed out. Jet planes shot over.

Tulpo came out of the main hut, leaning on a stick with both hands. Delores used a cane. He had kidney shunts, a common accessory here (a billboard outside Ajo offers discount dialysis). Gary's work to help bring back traditional food offers some hope for a future with less diabetes. For now the diet remains mixed. He had brought gifts of mesquite meal, fruit cocktail, canned corn, beans, Coke, Bisquick, and so on. The two old men sat with Cokes and bagels.

The hermit was about five-foot-five, with an undershot jaw, shoulder-length steel-gray hair, and a gentle, querulous smile. He was fascinated with my white beard, making shaving motions, twinkling and laughing softly — I wasn't entirely certain of the message, but we laughed together. He was also intrigued with my binoculars and gestured his desire to look through them. Remembering the wine-bottle monarch last night, I was a little concerned. These binos had been through a lot and were not replaceable. I asked Gary whether by O'odham custom, politeness required an object's admiration to be met with its gift. Gary, amused, said no, that wouldn't be a problem. I handed the binoculars to Tulpo and savored his distant view.

Two tribal members drove up in a truck, bringing groceries to Tulpo. They said the Fish and Wildlife Service brought water in a tanker. They helped Peter learn a little more about Tulpo's thoughts, and then it was time for us to go. I felt privileged to have had this glimpse of a rare man and his rarer way of life.

Gassing up at Why and waving the others off, I turned south toward Organ Pipe National Monument, heading for Quitobaquito Springs. In *The Desert Smells Like Rain,* Gary describes the springs, known as A'al Waipia by the Tohono O'odham, as a "modest pond touted as one of the few authentic desert oases on the continent." Until the National Park Service depopulated A'al Waipia in 1957, it had been occupied for a very long time by O'odham farmers practicing a nonintrusive kind of agriculture. Since then, the diversity of birds and plants has actually declined, compared to a traditional O'odham oasis across the border. Even so, Gary's description of the lush place had made me wonder if it might not attract monarchs to roost before passing farther south.

At the Organ Pipe visitor center, two young park rangers told me that the road to Quitobaquito was closed because of a landslide on the far side of a park loop road and a colony of Africanized bees at the oasis. I took my chances, turned right before the border, and drove past the orange "road closed" sign. After checking for aggressive bees, I rounded the pond. There were three remaining Frémont's cottonwoods, all female, their yellow leaves peppershot by moth larvae. The fallen, punched-out leaves surrounded an anthill fifteen feet across, its outer rim composed of soft golden seed chaff, like a shield volcano erupting duckling down. The pond was ringed by gray thorn, bulrush, a bit of willow. A green-backed heron watched from a dead branch, its red ruff, yellow legs and bill, chartreuse eye, and greeny crest all field-guide bright in the sun. And, of course, there were coots. A great egret fluttered up and around, alighting in a flurry of white plumes on the smaller heron's perch. Now the green-back walked on water, striding across pond scum lit up by red dragonflies. The egret flew again, its throat a sharp keel.

Sure, there was life here, as there would be at any desert waterhole. But I could imagine that there was even more life before, when the indigenous people stirred the mix through their careful, low-tech ways, as they had over the centuries. Wherever ancient agricultural patterns have been supplanted, local extinctions of animals and plants have followed.

This might seem heretical to conservationists who think nature is always richer in the absence of man. But British butterfly research has shown that where sustainable agriculture has coevolved with the native fauna and flora for millennia, the reverse can be true. Every instance of butterfly endangerment in Britain today can be rigorously linked to the disruption of old, established land uses. Twice the number of bird species had been recorded at the occupied and worked Mexican oasis than at this Park Service bird sanctuary, where the old houses had been bulldozed and the few remaining old pomegranates and figs languished among mesquite.

I saw no sign of killer bees, but the wash crossing under the border fence was full of scrambled human footprints. I didn't doubt that this was a crossing spot for illegal immigrants, who might hole up at the oasis for a while before attempting to move on. Gary wrote that drug runners in the area "sometimes shot anyone in sight — not just official-looking gringos, but unassuming Indians, too." As an unassuming gringo, I recognized that the formerly peaceful oasis was now a place of tension. I believed the road was closed not for any bees or slides but at the behest of the Border Patrol.

I returned to the Organ Pipe campground to spend the night. On the way back, as the sun set behind me, saguaros went red, the Ajo Mountains lavender. *El coyote* the hunter, *los conejos* the rabbits, and *los venados* the deer crossed my road, but only a flicker of what could have been *la monarca.*

Later, having fallen asleep on the picnic table, I awakened to the crescent moon, bright stars, cicadas singing, and lots of German chatter from a large group in rented campers. When they shut down, precisely at the quiet time of 10 P.M., the peace was complete. Orion rose over the organ pipes, crickets trilled, some night creature tapped at cactus pads. Closing my eyes, I saw again the ancient eyes of Tulpo and thought with a dull sadness of the O'odham, able to wander at will across a nonexistent frontier, from oasis to oasis. I imagined monarchs doing the same. I was so close to Mexico.

On a stump by the edge of the footpath in the forest of El Rosario, I am among the Mexican monarchs once more. Struggling in the dust, fluttering in the bushes, unwinged by orioles or disembodied by grosbeaks, but mostly whole and aloft or arrayed in the firs, they fill every space with their orange bustle and dun repose. Their forms — deltoid across the green and sunstruck distance — launch, drop, fall, rise, slip sideways through the terpene-sweet air. Mesmerized by the black flitter in the dust and its citrus shadows on the air, I perch as inert as the bundled butterflies hanging from the still-shadowed boughs. Over my own heart, I hear the soft whirr of millions of wings fanning the alpine air.

At Sierra Chincua, beyond the Llano de Papos, where I first saw the winter monarchs, the air seems filled with monarchs until a cloud covers the sun, the trees' burden explodes into flight, and the sky thickens with moving wings. The scents of oyamel *and* yerba buena *seem to ride on the satin rustle of countless wings. Sounds of Velcro, zippers, nylon, and clicking, buzzing cameras; soft voices; minutes of nothing but the breeze and the susurrus of fanning wings and fastening tarsi.*

Monarchs drop from the willowware sky onto tough-leaved, scrubby oaks, onto my limbs, my writing hand, my hot skin. The oak leaves shine red and green, the monarchs against them shimmer with an almost inner light as of ancient amber. Some have wings clipped by circumstance, truncated by encounter, chipped, bitten, torn, scratched, rubbed, or lost altogether. Others are perfect: I can make out their scale rows, as deep and rich as fox fur in February. The stirring, flickering shadows pass over hypnotically, like speeded-up cloudlets. And the sailing: overhead, out in front, down below against the dark fir wall, every possible direction every second.

Feinting, falling, swooping and soaring, flopping, flicking, hovering, floating; lighting, lifting, yawing, gliding; volplaning, parachuting, and stalling. Fine and swooping, like long Mexican lashes, butterfly tongues suck salt from denim on my knee and thigh, from my shoulder, from my sweated page. And as I lean back and cross my hands over my watering eyes to see the butterflies against the sun, a pale monarch alights on my fingers and drinks my very tears.

Morning in the cactus garden. I sat surrounded by big hook-spurred barrels, purple-fruited prickly pears, deceptively soft-furred chollas, massive saguaros, and the organ pipes themselves. The monument supports twenty-six species of cacti, including most of the organ pipes *(Stenocereus thurberi)* in the United States. On several trips I had tried to come to terms with saguaros *(Carnegiea gigantea)*, whose potent presence, I was glad to learn, was undimmed by a lifetime of comic-book versions and kitschy inflatable plastic cacti. But until this visit I had never seen the other species of monumental cactus, as they are called. To me, an organ pipe resembled a whole pack of slender saguaros growing from a common root. I felt I was in the presence of silent giants.

There was nothing silent about the avian company. At breakfast I was joined by two Gila woodpeckers, two curve-billed thrashers, two mourning doves, six Gambel's quails, and a feisty cactus wren, all walking or hopping around my table and pecking at spilled oats and each other, cherking, chuffling, wipping, chacking, and mewing. The wren foraged all through Powdermilk, just like the canyon wren at Glen Canyon. By eight o'clock the day was already hot, and I felt I had to leave the birds.

The other road I'd planned to take in Organ Pipe, the Camino de Dos Republicos, was closed off, too. Like the oasis road, it ran parallel to the border, but easterly. Again I ignored the "closed" sign and drove two miles to a place called Gachada Line Camp. The Park Service had made this a nice interpretive site, but the information signs were now uprooted. Raised, they told me this had been a Papago water source for hundreds of years. From 1916 to 1919 rancher Lon Blankenship dug a well to increase the flow for cattle. He sold the place to Robert Gray, whose headquarters was the Dos Lomitas Ranch, three miles east. For fifty years Gachada served as a remote base for Gray's cowboys.

The 1930s bunkhouse — now a neatly stuccoed adobe, slightly gone to seed — was originally cobbled together out of cardboard, creosote brush, cactus ribs, mud, road blacktop, and adobe bricks, rather like Tulpo's place. There was a strong tincture of bat urine as I looked in, and three big, handsome gray bats dropped out of the rafters and swooped into a flue at the bottom of the adobe chimney, silently. They were leaf-nosed bats, a species active throughout the year. Outside grew huge tamarisks. The old corral, long cattle-free, held only black-throated sparrows and thrashers. A cactus wren messed about between the corral's twisty posts and a cholla. A shadow by the adobe briefly became a monarch, then metamorphosed into an enormous saffron dragonfly.

In the sun, the temperature already near 90, I heard the high voices of Mexican children playing. I walked a few steps, planted my feet under the little barbed wire fence, and stood in Mexico. As a student of biogeography, I have always held borders in roughly the same degree of contempt as I do astrology and okra. Here were Mexican kids in a Mexican yard, Mexican roosters crowing, and Mexican cotton trucks chugging, all just a soccer field away from where my car was parked in the United States. Though I have family in Guadalajara, I agree with those who consider immigration a serious population issue for the United States. Yet on a gut level, borders make no sense to me at all. Monarchs, at least — like the coyote Thea and I had watched trotting across the Canadian border weeks before — display not the slightest obeisance to human lines on a map.

Returning, I saw that the mesquite arroyos crossing the Camino were pocked with footprints of javelinas and scored with tire tracks. Here and there lay a thrashed, abandoned truck tire. The washes of bare gravel and creosote bush made an easy route for "coyotes," those usurious agents for undocumented immigrants, and drug runners. I knew why these park roads were closed to sites that people ought to be able to visit. They were used by all sorts of migrants. Doubtless mine were among them.

FOURTEEN

Cibola

Leaving arizona, I crossed the border at a place the map called Lukeville, whose post office was labeled "Gringo Pass." The sign before the checkpoint read:

> Customs General Administration Warning:
> Introducing weapons, Buchshot or Explosives
> into mexico, without declaring them to customs
> authorities and without permission of the
> commerce department and national defe-
> nse its a crime that is sanction with
> a penalty of 2 to 30 years in prison.
> If you have weapons, Buchshot or
> Explosives, Declare them to custom
> authorities and surrender them Before
> ac-tivating the control Light passengers.

followed by pictures of various sinister guns. Butterfly nets were apparently exempted, so I did not declare Marsha.

At the border store, SUVs and ORVs were fueling up for Rocky Point, a resort on the Sea of Cortez. A well-paunched longhair with a three-wheeler stumbled over the doorway and grunted. His T-shirt proclaimed "What this country needs is a no-nonsense, take-charge kind of clown." His woman friend's extra-large asserted "Real Women Love Jesus." Earlier in the trip I'd heard of an organization that these two definitely ought to know about: "Clowns for Christ." Across the border lay Sonoita.

I did not intend to drive down to the monarchs' wintering areas in Michoacán nor try to follow monarchs there, but I did hope to see wanderers along the border, which Mexican Highway 2 closely follows all the way to the Colorado River. No one was able to tell me whether this route was a "free zone," like the road to Rocky Point — free of the need for a tourist card and special insurance. At first the road closely paralleled the Quitobaquito road on the American side. I passed a sign to Ejido Papago. In a draw full of rabbitbrush nectared cabbage whites, which had also been numerous at the oasis. "Of Cabbages & Coots" might have made another suitable motto for this trip.

A white Border Patrol Blazer drove alongside me, across the fence in the United States of America, headed toward the oasis where yesterday I had successfully evaded both killer bees and monarchs. Now I thought for a moment that I saw one, but it was a queen flying north to the border. Where the road turned a little south to climb low hills, another big pale queen gave me a run. I never thought I'd say this, but leaving their range would be almost a relief. The viceroys of Washington, the queens of Arizona, are delights in their own right, but there is much to be said for a land where a monarch is a monarch.

Pretty little poppies, two-tone orange, relieved the manufactured colors of the roadside verge. The road was my first asphalt washboard, and most of the traffic was fast funky Mexican trucks, fast fancy Mexican buses, and still faster Mexican cars. The posted speed limit of 60 km/hr was clearly a local joke. Few Americans seemed to use this shortcut west, perhaps preferring the security of I-8's divided highway, though Yuma was a hundred miles farther that way. Passing a wrecking yard full of smashed buses, as a bus passed me at great speed, *Radar en Operacion* notwithstanding, I could see their point. Elaborate monuments to the accidentally dead appeared along the roadside at frequent intervals. Within one mile: "Alejandro Heredia Mendoza." "Rafael Marcos Villaea 21 Marzo 78." "Mary García de Martínez 21 Marzo 78." "Arturo Velarde Ríos — Aug. 1, 91." Between the rushing trucks and buses, there was no time to view their sentiments and plastic flowers without becoming the latest one.

I pulled off for a nap in an *Area de Descanso,* with a lone palo verde by a basalt picnic table and a landfill's worth of trash. If litter along U.S. roadsides is legion, in Mexico it is regimental, and there are no adopt-a-highway programs. Tires, glass, and plastic lined every mile of the road. Farther on, near Ejido 5 de Mayo, I saw the prettiest litter: a load of pottery spilled from a wrecked or poorly packed truck. They were sunflower pots, of the beautiful blue, green, and yellow pattern called Talavera. All were shattered, but I salvaged the largest, most intact piece, half of a deep, round-handled bowl. The next spring it would be grown all through with bleeding heart, oxalis shamrocks, and maidenhair fern.

The air stood still at 93 F. The road ran on across a bristly cholla plain with cinder cones, sharp, extruded ranges, and rugged pink rock riddled with caves. The magnificent saguaros thinned out to clumps, then ones and twos. The radio brought mariachi from Mexicali and Sinatra from Yuma. Tiny settlements appeared, with a house or two in a big, bare gravel, stuff-strewn yard; a sign saying "*Llantero*" under shade with old tires and dismantled cars; a little ramshackle bull ring; and a café, a shed advertising "*deliciosa,*" Tecate, etc., and called Cupita or La Joyita. One such hamlet was named Minas de la Desierta, another El Sahuaro. At a truck stop with half a dozen trucks and two chapels in the middle of rocky nowhere, a bus stopped to let someone off. An old Indian walked along lugging a gallon of water and a bag of rice over his shoulder. I would have offered him a ride, but Powdermilk was stuffed. The Indian villages along the road looked mostly deserted, with brick or cinder-block houses never finished.

I was stopped at a *Sanitaria* checkpoint. A handsome woman in a black Policía T-shirt came out smiling to question me, but, not speaking English, she handed me over to a bilingual colleague, who asked where I was coming from and where I was going. As if I knew! My border route was not a common one for gringos, though it was, it turned out, a "free zone." The policeman smiled broadly and waved me on, saying, "Have a very good time!"

I passed Ejido de la Reforma Agraria, though there was little sign of anything agrarian, reformed or otherwise. In the Yuma Desert, practically the only plants I saw were creosote bush and a minute yellow aster. Then I was across that desert stretch into sand and cinder blocks on the outskirts of San Luis Río Colorado. A billboard, picturing a pretty green roadside, declared: "Area de Amortiguamiento Ecologico — Sedepro Parque Industriales Sonora." Place of Deadened Ecology — or of Ecological Softening? The scene more resembled the former. Suddenly I found myself in a town of 120,000 with many small houses, streets, shops, and garages. Parking near the green central plaza, I skipped the "Pub & Rock Caffe" and ducked into Kuki's Refresquería for a couple of taquitoes, hot and good. Then I picked up a couple of Tecates at an Oxo market and basked in the sunny plaza.

A woman in a tight green polka-dot skirt and nylons walked by with two girls in T-shirts and jeans, and an old vaquero got out of his truck and watched the curvy one out of sight. On the plaza, pigeons landed on people's heads beneath tall date palms. A remarkable hobo with dreadlocked beard and hair, almost black Indian skin, a plastic cowboy hat, greasy clothes, and bandoleers of ornaments and jewelry around his neck circled in a repetitive walk, meeting no other eyes with his black, faraway gaze. The schoolchildren walked past him without even noticing. I wondered how a person gets that way. I felt my matted hair, my sundark skin, my dirty jeans, my beaded necklace . . . maybe it was time to think about turning north.

I wanted to see where the Colorado River crossed the border, to see if it still looked like a monarch corridor. But heading west out of San Luis on the Mexicali road, I could see another police roadblock before the highway crossed the river. I felt I could never explain why I wanted to drive across the Colorado, then drive back again. So I turned around and aimed north. I would try to get a look at the river once I crossed back into the States.

In the long line for customs in San Luis, windshield washerboys messed up my glass for a few pesos, and vendors hawked Popsicles, green peppers, pastries, toys, balloons, puppets. Two customs lines bore different instructions that I could not read. I chose one, but all the Arizona plates seemed to be in the other, shorter line, so I pulled out, drove back, and went around. Then I saw U.S. plates in the first line, which was going faster. Near the checkpoint the two lines merged anyway, with no distinction. At customs, no one told me to have a good time.

The road north in Arizona crossed bright green fields of lettuce and black ones of cotton. I knew the Colorado River was nearby, but I couldn't seem to approach it. All the water was in fields being strafed by crop dusters. I continued north on a posted dike road between two canals. I asked a Mexican irrigator, "Dónde Rio Colorado?" He waved off westerly. "Dónde rio *vista*?" I asked. "Ah," he said. "No está aqui!"

In Yuma, electric colors of oleander, bougainvillea, and strip development lined the streets. I asked the fellow in Arizona Donuts where I might see the river. "It's pretty much the freeway bridge," he said. "Or you can walk along it by the old prison. But it's not much to look at." Driving down a forbidden road into a snowbird trailer park in North Yuma, I finally found a reedy, sluggish, bullfroggy backwater at Kokapah Bend. It smelled like a river, but the doughnut lad was right: it didn't look like much.

I left the private park on a levee that should have hit the river channel but instead swung around south and showed no signs of ever ending. I was halfway back to Mexico and the night was black when several sets of headlights flashed ahead on the levee. Approaching slowly, I saw three white-and-green Border Patrol Broncos blocking the dike road, agents lounging beside them. They stood straighter and watched, hands hovering near their holsters, as I stopped and doused my one operating headlight. I got out and began to approach them.

Holding my own hands in plain view in their headlights, I smiled. "Gentlemen," I called out, and tried to explain what I was doing there. "So you see, I'm trying to get an idea of what the Colorado River is like near the border, to see what kind of monarch butterfly habitat it might offer."

The officers looked me up and down and visibly relaxed. They seemed a little bored out there on the dike, and if I was an anticlimax, at least I was something different. I was half expecting a hot tip for a butterfly expressway, but the Border Patrol's biogeographical expertise must have been limited to their colleague back in Douglas.

One of them looked out into the night, spat, and said, "So you just wanta get a view of the Colorado River, huh?"

"Yes, sir," I nodded.

He exchanged looks with his partners, then drawled, "It can't be done."

Back in Yuma, I stopped at a Chevron for gas, coffee, and a Butterfinger. I crossed the Colorado at Yuma Crossing Historical Site, but the riverside park was closed, and I could see only glimmers of the river below. I was in California.

I turned north. All the way past Stud Mountain, Picacho Peak, and Quartz Peak on my right, the Imperial Valley, Giant Sand Dunes, and Chocolate Mountains on my left, I saw only palo verde and creosote in the headlights. Across the river the road led to "Arizona's West Coast," Lake Havasu City, home of the London Bridge and jetskiing championships. On my side, a fruit checkpoint and a Border Patrol roadblock looking for illegal immigrants; both waved me through.

Before midnight I found the Cibola National Wildlife Refuge, crossed its bridge and levee back into Arizona, and wandered backwater roads to a quiet place. Camping was not allowed, but I was out of steam and saw no alternative. The night was warm, starry, a yellow moon half-returned and dipping toward Ventura. The dirt roads were scattered with feral burro turds, and I finally saw a couple of shy donkeys slipping around a bend — the first night wildlife I'd seen all the way from Yuma. The hum of mosquitoes and the indifferent brays of burros were the final voices of the day.

I spent what was left of the night battling mosquitoes, but it was worth it to awaken to 70 degrees at seven, surrounded by Colorado River riparian and silence. Except for a pair of flickers, birds were remarkably quiet. Burros, then coyotes, then shotguns, a distant motor, a far plane. It was our anniversary, and I opened the card Thea had given me: "To share everything that nature has to offer." That sounded good. There had been plenty since we'd parted that she would have enjoyed sharing, some that she wouldn't. I'd been right to go it alone.

Creosote bush, I found, scoured breakfast dishes nearly as well as Russian olive, whose gritty foliage I had often used for that purpose. I poured my wash water on the bush in appreciation for its scouring sprig, and traded my own recycled fluids for another spray of the sticky twigs that I took for their powerful pungency.

By eight o'clock it was 80, and I walked out onto the wildlife refuge. It had been established in 1964 to protect riverine bottomlands "surrounded by a small fringe of desert ridges [like the one I was on] and washes." The dredged Colorado River channel, "built by the Bureau of Reclamation," bisected the refuge, while the old river course formed a marshy backwater. The gravelly upland supported mostly creosote bush, while most of the floodplain appeared to be a dusty bosque largely of tamarisk, with some mesquite. Boar, burro, and coyote tracks pocked the beaten track into the thicket by Cibola Lake. I crossed a deep-crust plain, a savannah owned by the burros, and climbed a flat-topped hill of cobbly slag. This stony terrace had been built from the spoils of dredging and hydraulic mining.

Ragged mountains topped out far off on all sides. I looked down onto reedy pools with lots of cormorants basking on snags and, of course, more coots. What I first took for western grebes turned out to be Clark's grebes, a new bird to me — white area around eye, bright yellow bill, *krick krick* call. Leopard frogs nattered. High piping emanated from the marsh, maybe from the endemic, endangered Yuma clapper rails that I knew were in there. Up where I was, a shrike rode a breeze, and a rufous-sided towhee scratched at the edge of the rocks. Pop! This part was closed to hunting, but the rest of "the refuge provides an extensive hunting program" (though you may not collect an insect). A funnel web slung from a little creosote bush was lined with seeds, like milkweed fluff stuck to the Invisible Man.

Pottery, glass, quartz, and every color of cobble dotted the slag. Pebbles and bottle caps were imbedded in an old chunk of concrete left over from a long-ago dwelling. Some sorts of lives were lived here, maybe by the bringers of the burros, even before the tamarisk arrived. Concrete foundations poked out of the northwest corner of the ridge, and on the east slope lay the dead bodies of '40s cars — a

Buick Eight, a Ford. Powdermilk waited nearby but, repelled by impenetrable tamarisk, I couldn't get there from here. I had to walk back to Burro Flats and around. Following a donkey path down the rocky slope, I could see why these animals, with their mincing little hooves, do so well in Grand Canyon. On the way, thirteen painted ladies passed me toward the southeast. Then, in the car, dust-surfing, I saw a stick- and mud-thatched, rude log cabin perched above Hart Mine Marsh. Finally, there lay the Colorado: the same river that cut Glen Canyon, channeled here but broad, blue, and beautiful all the same.

The canalized Colorado had good gravel roads on both sides — finally, a surfeit of access. I headed south along the hundred-year-levee road. Three young girls in swimsuits were walking along it, and a coyote dashed off. Except for girls and coyotes instead of camels and bee-eaters, the scene reminded me of the Kara Kum Canal in Turkmenia, where I had seen the Asian monarch among the tamarisk — right down to the tall *Phragmites* rushes, the sky bleached dusty-misty at the edges, and mountains of cotton in the near distance. But instead of *Danaus chrysippus,* I soon saw *Danaus gilippus* and then *Danaus plexippus.* A blooming desert broom was the first nectar of any kind I'd seen here other than the last pale pink tamarisk plumes and one palo verde in unseasonable yellow and green bloom. There was not a butterfly on the broom. At least there were no bloody queens to confuse the issue. But then a bigger clump showed a patch of orange. It was a queen after all, but I got out to look and spotted a big fresh monarch in the same bush. It was wonderful to see both American milkweed butterflies side by side on this desert canal.

I climbed down the rock embankment, but when I was near enough I took one extra step for better footing, and the monarch flew. I swooped and missed. It flew up fifty feet or so, then north along the dike some eighty yards, and parachuted down right onto the first *Baccharis.* Moving up, I couldn't see it, and figured it must have rounded the bush to the river side, where it might as well have been on the dark side of the moon.

This sighting was a mile upstream and across from Walter's Camp, an old fishing base. I carried on south to a point above Gilmore

Camp, where the dike road ended at a hand-driven cable car for crossing the water. Here the Colorado River broke out of the channel and ran wild for a distance. A jetski shrilled along, chasing away a great egret. The two boys tearing about on the water Yamaha were obviously having a good time, even if the egret was not. The young girls I had seen before were trying to get up on a board pulled by a motorboat. Later, as I walked along the levee above from broom to broom, putting the binoculars to my eyes for each butterfly, the girls' parents saw me and called their daughters back to camp.

I got down to a little beach where two canoes were tied up. A painted lady dallied on the broom, and a minute ceraunus blue flitted among the delightful pods of screwbean mesquite, which looked like caterpillars on a stem. These *tornillo* fruits told me that the river here had something of its original floral composition, that it wasn't all salt cedar and Russian olive. As Susan Tweit wrote about the screwbean's habitat in *Seasons in the Desert,* "over 90 percent of the Southwest's once-verdant, tangled *bosques* have been cleared, plowed, or dried up . . . during the past century." The peoples of the region exploited mesquites abundantly for starch; the soils used them for nitrogen fixation. The ceraunus blue would use it, too, laying her eggs among its leaves and buds.

More queens, and the big orange dragonflies. Do dragons mimic monarchs? They do a pretty darn good job in certain lights and attitudes, and they share potential predators with monarchs — here, for example, tropical kingbirds. Back at the original monarch site, the queen was still in residence, but no sign of the king. A zebra-tailed lizard skedaddled on high, skimpy legs, and I glimpsed a little coatimundi dashing into the brush. Most of the *Baccharis* grew below the bank (as it did on the other side), so it was hard to check. I crossed onto an island with lilac-hued asters on its dike. The tall, toothy mountains that lay to the north yesterday were now far south.

At the end of the island road, where the old course of the Colorado met the dug channel, a roadrunner flicked at the very edge, its tail as blue as a magpie's. A queen fought a rising wind over a muddy-water-weedy delta where round, ribbed clams and a little pointed snail lived. It felt almost like a real river, although the west side of the old river bed bore cheek-by-butt house trailers. This side was wild but had a bulldozed and burnt track through the tamarisk and sand to the end, near Walter's Camp. I walked it a little before three, chasing ruby-crowned kinglets and black phoebes along before me.

Admirals, ladies, queens, and snouts were all tippling at the *Baccharis* bar. I was about to turn back when I looked up into a broom and saw, backlit, that magic dark triangle that means "nectaring butterfly." It was rather small and darkish, and I figured it would be a queen. But it turned out to be a nice fresh female monarch attracted to the creamy blooms. I waited for her to come lower and then, showing consistency, bricked it. Marsha's binding had grown loose, and her rim twisted around the pole as it struck the stiff broom, but I couldn't blame her. The monarch took off ESE at 115 degrees, though I was able to watch her go only fifty yards or so before the tall brush blocked my view. She seemed to be picking up speed and altitude on this vector toward the main channel, but she might have crossed over and dropped down to nectar more. I walked on, hoping for a third strike.

Coati tracks, scat full of curlicue acacia fruits, and clam shells marked the sand and mud. A tawny and black mutillid wasp, as big as a small bumblebee, was going my way. Known as velvet ants, the wingless female mutillids look like the teddy bears of insects. Along the Snake River I had watched a smaller, reddish species of velvet ant indefatigably exploring every inch of a large hot sand dune for the holes dug by ground-nesting bees or wasps. Finding an active nest, she would piratize the young. She skittered across the clean sand at great speed as long as I could watch. One rufous species enjoys the common name "cow killer." Though hyperbolic, this suggests the powerful venom some velvet ants deliver. What a sting this mutillid must have! Yet it was so pretty and cuddly looking that I might well have picked it up if I had not recognized it.

The island was a sea of tamarisk except for one lone willow. With more willow, there might be viceroys, too. I couldn't help but wonder what the Lower Colorado was like before the dams, channels, and cities, the fish camps and jetskis, salt cedar, cotton, and lettuce, pipes and canals and water compacts. Gary Nabhan writes of the little tule-filled ditches that the squash- and bean-growing Indians made when the river was theirs. It seemed a suitable scale of irrigation, an appropriate demand to make on a river. But I was tired of feeling elegiac. Maybe because I first found nature along a beat-up ditch on farmed-out plains, I take solace and pleasure in the "unofficial countryside," as English naturalist Richard Mabey calls damaged places. I guess Cibola, a wildlife refuge, is official countryside, but it is definitely damaged. Still, I found monarchs there. The river keeps on giving, no matter how much we take or expect from it. But for how long?

I circled back on the high dike under an avenue of stressed cottonwoods and eucalyptus. The white-barked eucalypts shone and tossed. I knew I'd be seeing monarchs roosting in eucs soon at the coast, but it would be nice to see some here, too. In the 4:30 P.M. brightness and dusty wind, Vanessas — painted ladies, red admirals — were flying crazy among these tall trees. It was the last hour of sunshine, the wind shrilling. The Vanessas would dart, flit, drop, and bask on the dike road until I got near. They always seem to love the end of daylight, becoming most animated when other butterflies have already gone to roost. I thought I glimpsed a monarch, too, up on this windy levee, along with the ladies and the russet-breasted barn swallows swinging through.

Dove hunters arrived on the dike, looking unhappy that I was there first. Hearing lots of shots, I quit the avenue of poplars and gum trees after my second pass. In the last sun I tried a row of *Baccharis* along a canal west of the dike, across a sandy blowout. Then, hearing an unmistakable, haunting gabble, I looked up to see fifteen enormous sandhill cranes coming into Cibola from the northwest in three lots of five. Sandhills winter here. Were these just finishing their long migration from some distant northern breeding ground, maybe Malheur Lake in Oregon? Or would they carry on? Such a thing to witness, a long-distance migrant actually arriving at its terminus!

I have seen such a thing. On one trip to Mexico, I arrived before the bulk of the immigrating monarchs. The vanguard had already clumped into what looked like bushel basketfuls of butterflies in the tops of trees on Sierra Chincua's lower slopes. The monarchs would move up toward the Llano de los Lobos or down toward the Arroyo

Barranca Honda, depending on their reading of the forecast: higher if warm, lower if cold. Meanwhile, they continued to arrive by the thousands, pouring in like a heavy orange vapor. I sat on a rock, watching the train of them drifting up out of the valley below, over the rim of the mountain face, and into the forest of their winter shelter. I tried to imagine where each one had come from, how it had gotten here, what it had seen. But it became like trying to watch individual flecks of foam and water as they top a waterfall and plunge down into the foaming mass.

So I selected individuals to watch more closely. That was easy, as *Monarcas* landed on my hat, my boots, my hands. One lit on the edge of the page I was writing on, allowing me to draw it from life. Some of these butterflies had probably bred in Mexico and come from relatively nearby milkweed fields. Others had come hundreds, even thousands, of miles. I was actually observing the physical arrival of one of North America's premier migrants, coming home for the winter after weeks of strenuous, dangerous journeying down the continent. I shall never forget that sensation of privilege and awe.

Now, after the past few weeks traveling with monarchs, I felt I knew much more about how their journey passes. I'd seen them bed down, roost, and arise; seen them float, find a thermal, and lift out of sight; watched as they dropped to nectar and, sated (or scared by the intrusive observer), took off again. I'd observed dragonflies and swallows make passes at monarchs and either miss or think better of it. Seen the striped tubes and jade boxes from which each one comes, seen its mimics and its relatives. Above all, I'd seen and crossed the land that this so-called fragile creature transnavigates.

And the waters. Following monarchs, I had followed rivers, for they often go together. The first monarchs I saw as a child were winging along the banks of the High Line Canal from milkweed to goldenrod. As I prepared to leave the Colorado, I recalled that the river was one of the reasons I had set out on this trek in the first place: beholding that great vein on the map of the West, I knew it must be a vital conduit for monarchs in flow.

And so it is. Schoolteacher Mary Henshall's monarch, released in Boise, Idaho, and recovered in Orem, Utah, would (if it stayed on that track) have hit the Green River. There it would intersect the flyway where Mía Monroe, in September, had traveled with monarchs as she floated down Gray Canyon. About the same time, my friend Rick Brown watched another Utah monarch "heading south to the Colorado, dropping into the canyon northeast of Moab." Continuing, it would meet the paths of Mary's and Mía's monarchs at the confluence of the Green and the Colorado. And all of those, carrying on with the river, would come to the Grand Canyon, where Gary Nabhan had bobbed beneath a stream of them going his speed, his way. Maybe they left the river when it turned west, striking southeast for the Gila, the Salt, the San Pedro, the Guadalupe. But if they stuck with the river past Hoover Dam, past Las Vegas, they would eventually come to Cibola — where I found monarchs on the Colorado. My respect for both the rivers of water and the sky river of butterflies, if large before, was now profound.

Another writer who followed a wild migration, John Janovy, Jr., wrote in *Yellowlegs,* "What fortune, to fly with the yellowlegs, fly over the Gulf with your own wings; fly along the shore of Padre Island; fly through skimmers and laughing gulls; fly to Yucatan; watch the Brazilian jungle, Solomon's mine of hummingbird jewels, slowly pass beneath you, and every bit as tangible as the front yards, dogs, dripping cars and broken ride toys of a midwestern America passing before a walking scientist on leave! It is not so simple as you think, dirt-bound human. A migration with sandpipers would alter a person."

The creatures I was flying with never would see Brazil, the "Solomon's mine of hummingbird jewels," but here they were at Cibola. Cibola: the Seven Cities of Gold that Coronado sought on one of the truly epic wild-goose chases. There were no cities of gold, nor even golden geese, to be found across that sage and shadscale waste. But if aurelian powder counted for gold dust, then the treasure was right there beneath his eyes, beneath ours, all the time — spread across the continent on the wings of kings and queens in migration most majestic.

A migration with monarchs would alter a person, too. It has.

FIFTEEN

Pacifico

Winter was coming, and the butterflies were settling in. I needed to do the same. But before I headed home, I wanted to immerse myself in monarchs, and that meant going to the coast. Between here and there lay another desert — the Mojave — which I would only glimpse in passing.

Beyond the Colorado, Riverside County's evening found me amid more cotton, sorghum, pesticides, and a black field workers' shantytown, where a woman waved and children wobbled on old bikes. In Blythe I wished I could avail myself of the pretty pink Seashell Motel for eighteen dollars, but I wanted to get on. I stopped at the fruit checkpoint at Vidal Junction to ask for a highway map, surprising the state ag. inspection crew sawing open a coconut they had no doubt confiscated. They didn't have a decent map, just a joke of a tourist map with no detail. So I hauled out my old torn one from a time when a good map was every traveler's birthright. North past Grommet, Big, and Earp, under the Turtle Mountains, into night-pleasant Needles. Then onto I-40 west to Fenner Rest Area, between the Old Woman Mountains and the Providence Mountains, on the edge of the Mojave Desert. Which is where I found myself at the chilly dawning of the day, the 47-degree wind a shock after weeks of heat.

The Bonneville Salt Flats surprise still fresh in mind, I wondered if there might be any monarchs in *this* highway oasis? Work-release prisoners, men and women in variable bits of orange, were jumping around to keep warm. Cleaning out the trash bins, they salvaged aluminum and food. A gap-toothed blond with her orange cap on backward and a skinny guy in a Harley jacket got into a food fight. "Louie, you're going to hell for that." The birds were delighted with their leavings.

The bird fauna consisted of species equipped by natural selection for unnatural circumstances: English sparrow, Brewer's blackbird, rock dove, raven, grackle, mockingbird, half a dozen flickers, and, more remarkably, a Lewis's woodpecker. I pointed out the blue-black and rose woodpecker to a Texas woman taking Bitsy the Tabby and Timmy the Scottie on a bird walk. She'd seen my binos and told me she was a birder too. Lewis's was a lifer for her. "I've been looking for it for a long time," she said.

I asked the work-release woman if she ever saw butterflies coming out of the trees in the morning. She hadn't worked here much, but where she lived, by the river east of Needles, she saw them often, or so she thought. "They come in my garden window," she said. "Two nights ago there were twenty or forty in my closet. And you oughtta see my landlady's kingsnake," she continued. "Twelve feet long and" — making a boa-sized circle — "this big around." Later I saw her sitting in the sun, legs crossed, head down.

A burly trucker approached me and asked if I was watching birds. I told him I was following monarchs. "Yeah? You don't have to *follow* them or go nowhere to see 'em," he said. "Just look where that milkweed is!"

Into the Mojave Desert: creosote bush, which is absent from the Great Basin, and back to rabbitbrush, largely missing from the Sonora Desert. Straggly Joshua trees kicked in, a distinctive signature of the Mojave Desert. Fremont said of them: "Their stiff and ungraceful form makes them to the traveler the most repulsive trees in the vegetable kingdom." Fortunately not everyone thought so. Many years after Frémont came through, a national monument was set aside and named for these striking arboreal yuccas, where I once saw white-throated swifts copulating on the wing.

The line of semis, motor homes, and custom rods became a little thick, so when an exit sign appeared for "Historic 66: 11 miles," I took it. Writing in white stones stretched along the black embankments for miles — "KATY," "JESUS," "66." I stopped at Amboy, pop. 20, founded 1858. No inhabited place I have ever seen has looked more isolated. Cinder cone and black lava on the left, abysmal playa on the right. Motel, café, Amboy School. The school was still open — "Eight students, one principal, one nurse, two teachers, one bus driver," the French proprietor of the café/store told me as he served double-patty hamburgers to two very big persons: "Madame, Monsieur."

Back on the road, I was pressed on the left by a Burlington Northern/Santa Fe locomotive and on the right by a Harley Davidson. The skullcapped biker and I took the next mile of two-lane broken road together. Then he grew bored, peeled out to eighty between the faded yellow lines down the middle, and I had 66 to myself again. I passed a long hovel with a row of tall wooden packing boxes, each full of smaller cardboard boxes, and a great many rocks and horseshoes. Then shacks built onto railroad cars, old trailers, and carnival booths, and two Santa Fe boxcars with stovepipes joined by an arched ramada. Baghdad, California, was bombed out by time. Green finally reappeared in the first orchard, shoved into the caliche east of Barstow at Newberry Springs.

At one o'clock, east of Daggett, the first California monarch materialized from the air. A huge, bright one, it crossed Route 66 heading due south toward Daggett Pioneer Cemetery. I searched among a thicket of white wooden crosses, their names worn away by the desert wind. To have your very name taken by the wind! That seemed a good way to go, like the monarch itself, by then long gone south in the clutch of the wind. As a young boy, my favorite record was Gogi Grant's of "The Wayward Wind." It came back to me now, and I thought, if anything was born the next of kin to the wayward wind, it was this bound-to-wander butterfly. I searched dense rabbitbrush along a canal in line with the sighting, but there was no sign that anything so beautiful had blown that way.

Route 66 ran out at a Marine base with a cactus garden. I barely skimmed Barstow on a freeway, then fled back to the desert on Highway 58. Signs appeared for the 20-Mule Team Museum at Borax. In the early fifties, about when I was listening to Gogi Grant, my family first got TV. I was an avid viewer of *Death Valley Days,* hosted by Stanley Andrews as the Old Ranger long before Ronald Reagan got in on the act, and sponsored by Boraxo. I still like the soap, and I would sooner share the road with any number of twenty-mule teams than the twenty million trucks that clogged the right of way. Off to the edge, thousands of pale moths snowballed the rabbitbrush, and Joshua trees thickened into the northern Mojave.

Borax, then Boron, then Edwards Air Force Base — the western landing base for the space shuttles. As ravens crossed over going the other way, a monarch came flapping due south smack over the space shuttle landing strip. I watched it flap, soar, glide, orient, adjust, flap, dangle, and go, all driven by a brain the size of a poppy seed, fueled by flower nectar, bound from a distant somewhere in the North to who-knows-where in the South, right across the concreted desert floor where that big lunker of an aircraft flops down when the weather is bad in Florida, having depleted the ozone by a measurable amount with each lunge out and back. With all due respect for NASA, the astronauts, and all those rocket scientists at Morton Thiokol, I merely gawk at their handicrafts; but I am truly moved by these other fliers with their thin orange membranes, their delicate toughness. And as for human technological magnificence? Just beyond the town of Mojave, to the dramatic climax of Beethoven's *Eroica* symphony, Powdermilk reached 250,000 miles. A quarter-million miles we'd gone together — to the moon!

The first miles on our way to half a million struck northwesterly into the Tehachapis, where a battalion of great windmills on a hill made it look about to take off. These spinners are hard on migrating birds, and I wondered if butterflies ever got caught up in their vortices. Maybe their natural aversion to wind keeps them down when the mills are spinning. Yet that aversion is not fully shared by migrating monarchs, which often fly directly into a headwind. At Tehachapi Summit, 4,064 feet, a notice cautioned "Reduce Speed When Wet." Better yet, I thought, drive dry.

Down then, dramatically into oaks and pines after the desert. The hills were like golden rolling cloud banks, the oaks like black ones. Then came the great green fertile expansive Central Valley, as the Joads saw it — imagine their relief! — in *The Grapes of Wrath,* which I'd been reading throughout the trip. I passed Weedpatch Highway, leading to where they were now in the story, the federal camp in Weedpatch. The dustbowl diaspora they belonged to was one of the great human movements of the century, inspired by ecological change, and no list of the literature of migration would be complete without it.

I aimed to give Bakersfield the slip nearly as quickly as Barstow, though it seemed to be a town with no directions out of town. Beyond I-5, the fifth California interstate I'd seen that day, a pleasingly diminished 58 bore westerly to Buttonwillow. "Heart of Cotton Country" said the sign, among giant bales of cotton and cotton fluff all along the road. I plucked a boll and it smelled sweet, like fresh dead leaves. "Food Grows Where Water Flows," read another sign, and the California Aqueduct flowed by in a barren concrete runnel, fluming Feather River water toward Babylon.

Over the Temblor Range, winding up and down and up into a cotton-candy sunset, then more hills into the California Valley. Another coast range to go, and no one else on the road but a lone cross-country bicyclist. It was cold. I hadn't been cold in a long time. At 8 P.M., I made U.S. 101, and struck the coast at Morro Bay. Ending a long drought, I bought a six-pack of Mendocino Redtail Ale, drank one to the sea and another to the moon, and bedded down — right beneath monarchs. Or where the state park map said they should be, anyway. I could not make them out in the flashlight or moonlight, but as I reclined and stretched out on sweet-smelling blue gum leaves, I knew I was in their presence.

At first light, I make out dark clumps in the tall eucalypts above me, but I'm not sure if they are monarchs or just leaves. Blackbirds and a western gray squirrel work over my campsite, fruitlessly. I watch the squirrel climb an enormous gum tree's trunk, out onto a branch, and then to a slender limb. As it leaps onto an adjoining limb, it triggers an explosion of golden particles into the first sunbeams. And in that instant I see more monarchs than I have in these past two months.

Soon another cluster, just behind and to the right, twenty yards from my waking face, begins to shed scales into the striking light — each scale a four-winged flier covered with smaller scales. Then bundles all over the grove devolve into falling, lifting, gliding monarch butterflies. The squirrel that started it all lopes about with others, pulling their splendid silver plumes behind them as they dart behind the ragged-barked trunks.

Rising and looking around, I saw that the monarch colony seemed loosely organized around one stupendous old two-trunked eucalyptus, seven or eight feet in diameter, with several clusters in it. Most were thirty to fifty feet up in the trees and ranged in size from punching bags to duffel bags. As the temperature and the sun climbed, and more monarchs took wing, their shadows dusted the ground and its crazy pavement of gum leaves. Rose, lime, and buttered toast, the leaves fluttered as they fell, their eccentric crescents twisting down, taking more than a minute to reach the ground.

Near me were just a few quiet campers, mostly tenters, in the shade of the big eucs. The smells of dust, bacon, campfire smoke, and eucalyptus oils mingled on the air. I took my own breakfast in the sun immediately beneath monarchs. Munching, I considered the massive aggregation of protein and fat all around me. Brewer's blackbirds, flying among them, ignored the monarchs, as they are supposed to. Do the parents teach the young, or does each have to learn the bitter lesson for itself? One butterfly had a big chunk snapped out of the left hindwing; another, the right forewing. Monarchs differ in their toxicity (scientists speak of the "palatability spectrum"), and some birds may be able to pick and choose by pecking and tasting, rejecting the bitterer butterflies.

Lincoln Brower has shown in the Mexican clusters that orioles slit open the abdomens to remove the tasty, nutritious bits. I once beheld scores of orioles of four species harvesting the monarchs of Rosario like porpoises skimming off a school of herring, as I stood in a rain of wings. In California, Walter Sakai has recorded black-headed grosbeaks, Scott's orioles, black-backed orioles, European starlings, chestnut-backed chickadees, scrub jays, spotted towhees, and western kingbirds feeding on monarchs, as well as deer mice, wasps, and ants, plus garden slugs feeding on dead ones. Again, when you have this much food in one place, *someone* will learn to exploit it. Now titmice were picking among the boughs. Anna's hummingbirds, no threat, flashed past a waking bunch.

A bohemian couple my age made their quiet breakfast in the cool shadows, the woman clad in a wool serape and beret. Why didn't they come out in the sun with us? Monarchs were beginning to bask. I went from wool to sweats to skin, allowing the autumn sunshine to fall on my bare back, basking with the butterflies: a poikilotherm like them, able to raise my temperature only so much through my own shivering.

As I lay back in close company with monarchs, a lyric from the Beach Boys' "California Saga" ran through my mind: "Have you ever been north of Morro Bay? The south coast plows the sea . . . There the monarch's autumn journey ends, on a windblown cypress tree." Many other poets have plumbed the heart of the beast in their lines, describing monarchs memorably as "an airborne chapel of levitating light" (James Worley), "a tide of fluttering orange and black" (Alison Deming), "El aire, como un rio llevaba corrientes de mariposas" (Homero Aridjis), and "your own inner blazonry" (Galway Kinnell). Novelists, too, such as John Updike in *The Witches of Eastwick:* "Brenda bent her head, and her mouth gave birth to an especially vivid, furry, foul-tasting monarch butterfly, its orange wings rimmed thickly in black, its flickering flight casual and indolent beneath the white-painted rafters." And Nabokov, who got their dangling lilt just right in *Pnin:* "Again, on serene afternoons, huge, amber-brown Monarch butterflies flapped over asphalt and lawn as they lazily drifted south, their incompletely retracted black legs hanging rather low beneath their polka-dotted bodies."

John Steinbeck commemorated monarchs in "Hooptedoodle (2), or the Pacific Grove Butterfly Festival," a chapter of *Sweet Thursday,* the sequel to *Cannery Row:* "On a certain day in the shouting springtime great clouds of orangy monarch butterflies, like twinkling aery fields of flowers, sail high in the air on a majestic pilgrimage across Monterey Bay and land in the outskirts of Pacific Grove in the pine woods. The butterflies know exactly where they are going." If you go there today, you will see a splendid (and remarkably airy) granite sculpture of *Danaus plexippus* by Gordon Newell on the waterfront, and another in bronze in front of the post office. "Butterfly Kids" by Christopher Bell depicts the "welcome back the monarch" parade, an annual tradition since 1939. Nearby, a large and resplendent stained glass window gracing an arch above the entrance to the restored Ford/Holman building depicts, of course, monarchs.

Numerous painters have portrayed the creature, and photographers such as Franz Lanting and George Lepp have captured it emphatically and gloriously on film. Weaver Nancy Klos Smith of Portland and printmaker Robert Johnson of Minneapolis have brilliantly illuminated the patterns of the adult's wing and the caterpillar's skin, respectively. I've enjoyed all of these expressions of essential danaid beauty. But here, sunning among them, I knew that the truest, most memorable expression of these "brightly fluttering bits of life," as Rachel Carson called them, resided in their own being.

Monarchs ornamenting a Monterey cypress bough like Czech glass birds clipped to a Christmas tree. A monarch two-thirds spread on a quartet of dried tan eucalyptus leaves, a sumi painting. Monarchs on a green-leaf crescent, one closed, one pumping, one wing-deformed. A monarch joining the one in the sumi — a fresh rendition. Another monarch folded against lime lichens on a cypress branch next to a cluster of hoary cones. Monarchs fluttering against the ragged, reticulated, flapping-in-the-breeze bark strips of the Grandmother Gum. Half a dozen monarchs jockeying to get on a spray of yellow leaves like a sheaf of sickles. Two monarchs with a forewing outside the hindwing — did they fly here that way? Can they get it back in? Monarchs on the wing, flap flap flap float, their black and white dots discrete, disembodied, the orange again lost against the background.

It was a kind of nirvana. I remembered this feeling from a spring morning in Mexico in the Arroya Barranca Honda, on the south slope of Sierra Chincua. As we climbed down a steep hillside to the late-season cluster site, known then simply as Julia, the sharp punchy scent of yerba buena filled my high-altitude-pumping lungs like menthol. A few monarchs floated here and there. It was hard to imagine that a major winter colony was near, but I'd seen the rusty smear across the forest wall from a distant vantage. Suddenly — and it was sudden — I passed through a scrim of firs known as *Abies religiosa,* or *oyamel,* and a second curtain hung in the sun before me: a skein of shimmering, moving sequins. From the floor of the canyon to the tops of high trees well up the walls, every surface was glazed with a patina of bright wings.

Broad leaves bowed under the butterflies' weight, and the yellow and purple senecios and scarlet salvias were each attended by monarchs who had arrived with too little fat under their belts for the coming cold nights. Then, as now, I simply stood back and opened myself to everything around me, purely dominated by this one perfect organism and the soft sound of its millions of wings in flight, until all too soon the time came to part the curtain, leave Julia behind, and begin the steep climb back toward camp. Here at Morro Bay there were probably no more than ten thousand monarchs, compared to the millions at Julia or El Rosario. But sheer numbers become relative after a certain point of saturation. One can absorb only so much.

A fresh landward breeze lofted off the ocean, only a couple of hundred yards away. At 10:30, the temperature was still only 66. Monarchs were cruising near the ground, as I'd seldom seen them do during the migration. I wondered if what might be called their "attitude" (their relation to their surroundings, and how they act as a consequence) changed with the loss of urgency — with the con*tent,* you could say — of finally getting where they were bound. It must. Mine certainly did.

Surprisingly, monarchs do not appear in any known accounts of the Spanish/Mexican era in California history. In fact, there are no written records of them prior to an anonymous article published in 1874. Lucia Shepardson of Pacific Grove, who had been watching monarchs since 1898, published in 1914 a booklet entitled *The Butterfly*

Trees. She said they "rested thickly" on pine trees and came out on warm days to "play like birds." Correctly, she surmised that many came from other parts of California, and some from farther away. The foremost California monarch authority today, John Lane, thinks the original roost trees were chiefly Monterey pines and cypresses, coast redwoods, and other native conifers. As they were cut down and their groves fragmented, Australian eucalypts (chiefly blue gum, *E. globulus*) were brought in for timber and became widely established. The butterflies made a fortunate adaptive shift in their roosting trees. The leaves of the gums proved well shaped for perching in clusters, and the structure of the groves provides favorable conditions. Had the gum trees proved unsuitable or the monarchs less adaptable, the Californian monarch phenomenon might have faded out long ago.

Many people still believe that all of the western monarchs go to the Monterey Peninsula. Pacific Grove in particular is widely known as "Butterfly Town, U.S.A.," and an annual parade features children in monarch costumes. In 1938 an ordinance was passed protecting the butterflies (but not their trees) from molestation at penalty of a $500 fine. But the law did not prevent much of the habitat from being ruined — partly through development of facilities built to attract more monarch-watching tourists. In 1992, however, Friends of the Monarchs, under the leadership of Ro Vaccaro, convinced the citizens of Pacific Grove to pass a bond issue to acquire the famous "Butterfly Trees" behind the Butterfly Grove Inn and restore nearby degraded habitat. It is too soon to say whether the depleted clusters there will return in force. Lately the monarchs have largely abandoned Washington Park, another traditional Pacific Grove site whose trees have aged, but they have turned up on previously unknown, privately owned groves. The long-term future of the monarchs of "Butterfly Town, U.S.A." is controversial and far from certain.

But monarchs range far north and south of Monterey. More or less regular winter sites are found up and down much of the coast, from Bolinas in Marin County to Ensenada in Baja California, with some of the largest sites in Santa Cruz and Santa Barbara. Scientists working through the Monarch Project of the Xerces Society conducted a coastwide survey of cluster sites to prevent losses caused by lack of awareness of the monarchs' whereabouts. The society, led by Melody Mackey Allen, lobbied for the passage of Proposition 70, which allocated $2 million for monarch habitat acquisition. Later, Xerces' Katrin Snow (whom I had heard on the radio that dark night crossing Utah) prepared a monarch management guide for landowners. The sites became a part of the California Natural Diversity Data Base and are now monitored by members of the Monarch Program of San Diego. The program's director, David Marriott, works with volunteers, ranging from kids and moms to Professor Walter Sakai of Santa Monica College and Muir Woods National Monument superintendent Mía Monroe, to count and tag monarchs in winter sites up and down the coast, track their dispersal in the spring, and teach Californians about their most famous butterfly.

In 1976, the International Union for Conservation of Nature and Natural Resources (IUCN) declared the migratory monarchs of North America to be the number-one priority in world butterfly conservation. In 1983, IUCN's *Invertebrate Red Data Book* listed both the Mexican and Californian populations as Threatened Phenomena, a new category for species at risk, not of extinction but of depletion in their most dramatic occurrences. The idea of an endangered natural phenomenon, as applied to monarchs, was developed independently by Lincoln Brower and myself. Now, immersed in the phenomenon itself, I was thinking about the protection of monarchs when the silence of the Morro Bay grove was usurped by a mechanical wood chipper's roar. It has to be done sometime in a campground; the eucs shed a lot of biomass in bark and branches. But park workers tend to thin living limbs as well, and monarch specialists have a lot of concern over how much this can be done without harming the groves. Even more controversial is removal and replacement of eucalyptus groves throughout the state. The policy of California state parks, The Nature Conservancy, and other reserve authorities has been to remove exotic trees such as Australian eucalypts. They, and members of the California Native Plant Society, have been galled to hear that the gum trees are now considered important for monarchs and must be kept — even replanted — in some places.

These exotic trees have established their own ecosystem now, even if it is not the one that came before. Juncos work the leaf litter, brown creepers the trunks, chestnut-backed chickadees the foliage, lots of birds the canopy high up, and the native gray squirrels the whole place. We all love and revere native plants. But if Californians want to perpetuate Steinbeck's "great clouds of orangy monarch butterflies," they will have to make peace with these Australian trees that so dominate their woods now. After all, the eucalypts came to the rescue of the monarchs after Californians depleted the original roost trees for their own needs. It seems we must accommodate them, just as monarchs have accommodated to the gum trees, both here and in Australia, where they have evolved a distinctive wet season–dry season migration in the century they've been there.

Fluttering about the park on the breeze, some monarchs seemed to be looking for nectar. There wasn't much here, but later I saw a monarch nectaring on an exotic ice plant at the entrance. Most of the butterflies still looked quite fresh; only a few were worn, a surprising fact given that many had probably flown a significant distance to get there. A male and female perched side by side on a green gum leaf, civil, oblivious to each other. I saw a little one-on-one action, but very little, half-hearted, their hormones winter-damped. Come January and February, this would change dramatically.

C. V. Riley, a government entomologist of the last century who made the first careful studies and prescient speculations about monarch migration, was also a poet:

> Lazily flying
> Over the flower-decked prairies, West;
> Basking in sunshine till daylight is dying,
> And resting all night on Asclepias' breast;
> Joyously dancing,
> Merrily prancing
> Chasing his lady-love in the air,
> Fluttering gaily,
> Frolicking daily,
> Free from anxiety, sorrow, and care!

Riley brought a good deal more romanticism to the act of monarch mating than perhaps it deserves. In fact, no butterfly has a less ceremonious, more abrupt courtship than the monarch. I suppose "Chasing his lady-love in the air" would be one way of putting it. The male simply attacks the female on the wing, drives her to the ground, and wrestles with her. He will maneuver the female onto her *back,* wings spread, and cover her — a face-to-face embrace I've never seen among other butterflies. In a couple of minutes he will achieve copulation by enfolding the tip of her abdomen within the handlike claspers of his own rear end, and inserting his aedeagus. Then he will fly straight up, carrying her in a postnuptial flight, while she remains closed and inert, into a tree. There they will remain in coitus for an hour, two, or all night long, while he passes his seed packet (the spermatophore) to her *bursa copulatrix.*

Males mate as often as they are able, ten and sometimes twenty times, attempting to ensure the successful passing on of their genes. Unlike some butterflies, female monarchs mate repeatedly as well. This can increase their fecundity with fresher sperm; they absorb the redundant spermatophores as a form of nutrition for the protein and starch they carry. Females have even, rarely, been known to burst from too many spermatophores. Later matings, after the vernal frenzy, may be more deliberate, with males less aggressive and females choosing mates by their appearance and pheromones. But the "courtship" following the winter dormancy can only be considered ravishment.

By eleven, the air of the park was liberally sprinkled with the tangerine dream, yet apparently unnoticed except by me. This was a mystery, since people travel from far and wide to view the well-publicized and interpreted monarch groves of Pismo Beach or Natural Bridges State Park. But here at Morro, I did not see one other person — camper, walker, jogger, golfer on the adjacent course, or even birder (and there were a few around) — even look up. My neighbors vividly demonstrated Nabokov's lament about how few people notice butterflies — even in their thousands.

The squirrels got the last of my Rye Crisp. Each campsite was equipped with a wooden cabinet for a squirrel-proof food safe, but I was using mine as a desk. I opened a can of pears, exciting the yellowjackets, ate the pears, and left the can for them. Soon a gray squirrel had half its body inside the pear can, then it cleaned up the spilled oatmeal. Certainly it is one of the handsomest of American squirrels, with its agouti-gray back, cottony chest, vast plumose tail, and big ears, rivaling even the elegant Kaibabs. It is endangered in Washington within its shrinking oak-land habitat, but it's common here. As lumbering as they are attractive, Western gray squirrels end up far too frequently as road kills.

I remembered the ashen leftovers of *Sciurus griseus* lying about the roads of Sequoia National Park during a summer of rangering there. I also recalled monarchs, worn and rising in the High Sierra when I arrived in May, fresh and drifting back down to the lowlands when I left in September. Maybe some of them had come here, flying into gentle offshore winds until they hit the coast. This is how the autumn assembly comes together, from all over California and some ways beyond. Whether individuals "aim" for particular parts of the coast is unknown.

Now the onshore wind freshened, scattering monarchs, blowing one into my face. There was no apparent pattern to who flew off when or who settled where. Unlike bats, who return to the same spot in the cave, with a particular niche and a baby, the monarchs assemble, disassemble, and reassemble, like decks of cards constantly reshuffled in the air. They hung in a ball from a vine over the road, like a piñata. They glided, spiraling in to alight, setting others aflutter. They sat counterposed on a twig, like reflections in a still pool. Two of them pirouetted through the sky together as they fell. Many lined up across boughs like bright garments hanging on a laundry line.

On warm days Mexican monarchs come down to streamsides to take water. During one visit to El Rosario, I watched thousands sipping urgently by the trailside at a muddy spring. On another trek to a high meadow above Angangueo, I saw hundreds floating spread-winged on a puddle, drowned. It seems monarchs sometimes don't know when to stop drinking, become bloated, and cannot take flight again before night falls. Then, separated from the shelter of the cluster, they may freeze. Mortality is very high, of course, among virtually all insects. Migration exposes monarchs to hazards that sedentary butterflies don't have to face; many freeze in bad weather, many drown, and many fall to predators that key in on the winter masses. Yet monarchs are tough, and a surprising number survive the risky trip and make it through the winter, enough to repopulate the North. Still, it is poignant to see butterflies rime-coated on the forest floor or floating drowned, like water lilies. California monarchs are thought to drink dew here in the coastal fog belt where freezing is not a threat. Now I watched them skimming the ground, ignoring puddles around water pipes.

People are often surprised that monarchs do not migrate to the warm parts of Mexico. The point is to remain cool and torpid, not hot, which would cause them to burn their precious fat. Both the montane fir forests of Michoacán and this cool coast offer suitable climates and groves for overwintering. Few other places have the right combination, or the phenomenon would be more widespread. Now, with the temperature up to 75 degrees and the sea breeze brisk, many monarchs spread their wings in the sun; those in the shade were still close-winged. Even with many on the wing, I got the impression that most were bundled and more or less quiet — which is, after all, how monarchs are supposed to spend the winter. But when the shifting sun struck a hitherto shadowed branch, monarchs newly filled the air.

I saw no tags, left forewing or otherwise. They'd be easy to see, like flying billboards. Nor did I tag monarchs, lacking a permit to do so in the park. Reluctantly, I packed up and cleared out, stopping to pay seventeen dollars for my campsite beneath the roost trees — the most I'd spent anywhere, but a bargain here. I don't think camping should be allowed near the butterflies, but in the meantime, I was the happy beneficiary of bad policy. As always, leaving was a form of torture.

The ranger, a young woman, told me I was lucky, that they usually don't come there until January and February and that they often roost in a different part of the park. That surprised me, since January/February is when monarchs *leave* most California sites, much earlier than in Mexico. When we returned to Morro three months later, in January, Thea and I found the same trees still full of monarchs; but then the campground filled up with holiday campers and their smoke, and that night all the eucalyptus and cypress boughs were deserted.

I had no doubt the ranger was right: I *was* lucky. I had known here something of what my naturalist friend Mike Houck knew once, bicycling the coast:

> We had just abandoned Hwy 101 in the Redwoods and headed over the much loathed, but over-hyped "Leggett Hill" to join CA 1 at its northern terminus, just north of Fort Bragg. I cannot recall exactly where on CA 1 we saw the monarchs, but we cruised alongside small "flocks" of the butterflies as they wended their way south along the same course we were taking. Although it was a decade ago, I still remember the exhilaration of the warm October afternoons with monarchs wafting by, seemingly carried by on the steady breeze that also helped us on our way to San Francisco. In fact, I am sure that the migrating monarchs were a major contributor to one of the few "vaporized" moments in my life. I was standing on my pedals speeding down an extremely sinuous stretch of the highway, tracing the white line at the road's edge above the incredibly blue Pacific, when I got one of those feelings . . . "This is it! I've lived, let me vaporize now so this feeling will never go away."

On a beach below the natural history museum, I watched brown pelicans soaring offshore, ready to dive. Peeps herded a score of marbled godwits along the shore; Thayer's gulls and an immature ring-billed gull solicited a clammer for scraps. Later, in the little seaside town of Morro Bay, I sat in the sun with a shrimp pocket, a Russian tea biscuit, and strong dark coffee, considering how to go home.

My attempt to retreat from the coast foiled by U.S. 1, I landed accidentally in the village of Cayucos. On a steep road high on the town's east slope, I passed a garden full of purple lantana brimming with monarchs. Trying to look unobtrusive, I began catching monarchs with my forceps and tagging them. I could have done more with Marsha, but tweezing is gentler, far less conspicuous, fun, and probably easier on the flowers. A pretty young mother in a flowered dress, carrying a baby boy in diapers, came out to see what I was doing. After watching for a while, Tammy helped me with one female that had a broken costa, the stiff leading edge that is vital for strong flight. To our mutual delight, I was able to splint it with tag #81999. After 5 P.M. the air cooled and the garden quickly cleared out. Now there was just one last monarch, nectaring deep and intently on an orange blossom. She would be #82000, the last tag on the paper strip. I reached toward her with the stamp tongs. Mesmerized by the scent of the white flower, the butterfly's pale oval of citrus silk, I moved too slowly and the monarch flew off — the last of the day, but not of the journey.

When we returned in January, Tammy told us that monarchs had been around all winter, but she hadn't seen any of the tagged ones. The white flower had become an orange, and a west coast lady basked on a brick. There were no monarchs on her purple lantana. We did find a colony along flooded Cayucos Creek, however, with clusters in the crowns of Monterey pines, Monterey cypresses, and eucs, over nasturtiums, vinca, and the lemony yellow oxalis that blooms everywhere in coastal California's spring. Pussy willows were in flower, and hundreds of butterflies were flying, mating, basking. A red admiral chased a small female monarch, on whose crisp, carroty wing I finally applied tag #82000.

SIXTEEN

Fandango

I WAS TEMPTED TO TRAVEL home up the coast, watching for latecomers along the Mendocino bluffs and making courtesy calls at some of that region's estimable small breweries. I called my friend Mía Monroe, at Muir Woods, who described a spring mating frenzy at the Green Gulch Zen Center's milkweed and *Tithonia* garden and invited me to visit the Bolinas swarms. But I decided instead to cross the Sierra into Nevada and head north to the big lakes on the Oregon/California border in hopes of finding any last stragglers still migrating out of the inland Northwest.

Leaving the coast by a curvy tunnel through oaks and sycamores, I braked for a gray squirrel that someone else had not slowed for, felt its fine pelt, and placed it aside on a carpet of oak and poison oak leaves. Traffic thickened through Atascadero, Paso Robles, and Steinbeck's Salinas, becoming turgid by San Jose. In Campbell I slept in a bed for the first time since Tucson, passing the night and next morning with my sister Susan, her husband, Ted, and their four cats. There was a spray of ornamental eucalyptus leaves in my bathroom, and when their strong scent came to me, my mind's eye saw monarchs.

Next day, escaping the lowlands, I turned off at the Wente Brothers Winery in Livermore to see if there were any monarchs in the elaborate gardens. There were not. But in the tasting room I found I still enjoyed their Grey Riesling, a buck-fifty favorite in college, now four-fifty. And outside, white-throated swifts wheeled, climbed, stooped, and screamed in their ecstatic flight above the water tank on the hill — after all the classic canyons where they'd been absent. Two Latino gardeners noticed me eyeballing the swifts, and one said, "If they're migrating, maybe they stop here to have a drink of wine, eh?"

"Yeah," I said, "like me." California sister butterflies are known to tipple immoderately at the spilled ullage of wineries, and I've watched buckeyes doing so at Ravenswood Winery in Napa. It is not inconceivable that monarchs do the same.

Before leaving Livermore I walked the Arroyo Mocho trail, where a different kind of buckeye dominated — the tree, not the butterfly. The big, burnished fruits of the California buckeye lay about, shining like oxblood Oxfords polished for years. California sycamores lined the gulch, a sometime roost tree for monarchs in southern Californian sites. Here none were in evidence, just sooty-wing and fiery skippers on white sweet clover, and fields of fennel where anise swallowtails would fly in spring.

Visibility was poor in the Central Valley due to field burning. On the radio I learned that brush and houses were burning at Malibu and Big Sur. Had I chosen to travel up the coast, I would have found U.S. 1 closed to the north by the wildfires. Outside Tracy, dairy land was bracketed between Pombo Real Estate signs offering it for high-density development and others offering "Richard Pombo for Congress — he's one of us." Near Farmington I passed fields of pumpkins, walnuts, and tomatoes. Ground squirrels bellied up to the warm curbstones, like preachers at their altars. When I was across I-5, this time to the east, and on a small country road, the thousands of people became dozens, then none, and it was better.

Signs pointed to Lodi — where I was honest-to-God stuck once, in a two-dollar hotel with bedbugs — and to Copperopolis, just as the radio broadcast a story on copper trading. Traffic came back at Angel's Camp, the home of Mark Twain's Calaveras County frog-jumping tale, on the main line from the capital to Yosemite. Little Route 4 climbed into Sierra Nevada foothills more developed than I'd expected, with strip shoppettes, chain stores, and a country club. The bright side was the Snowshoe Brewery, where the World Series commanded the obligatory big-screen TV fixed on ESPN. The ESB (extra special bitter), with five types of hops, was of much more interest to me. But the excellent ale was not meant for motoring. I camped across the road in the Calaveras Grove of the Big Trees. Wrapped deep in my bag against the mountain cold, I slept until 9:30 the next morning, when the tinkle of golden-crowned kinglets woke me.

With freezing fingers I washed and diced onion, tomato, and long red peppers, fried bread and an omelet, and made tea, to the cussing of Steller's jays, nuthatches, chickadees, and chickarees. A gray-mustached ranger collected my twelve dollars, leaving me with a few cents in my pockets along with my gasoline credit card. Sun fell on broad red dogwood leaves. Incense cedars, sugar pines, grand firs, and Big Trees shook off their frost.

I hadn't been among giant sequoias since my summer as a ranger-naturalist in Sequoia National Park, twenty-seven years earlier. Of this grove, John Muir wrote: "They were the first discovered and are the best known. Thousands of travelers from every country have come to pay them tribute of admiration and praise." But he railed bitterly about the treatment of one tree, now a burned-out snag, at the end of the Big Tree Trail: 200 feet high, with many hollows, one arm raised near the top, and one arm out lower down. I walked down to have a look at what's left of it.

The park brochure called it "the sacrificial tree," though it was once known as "Mother of the Forest." Its bark was stripped, shipped, and reassembled for displays in New York City and London in 1854. Muir referred to it as a "ghastly disfigured ruin" and protested that skinning this great tree alive was "as sensible a scheme as skinning our great men would be to prove their greatness." A sixpence brochure from the time when the "Mammoth Tree of California" was erected at the Crystal Palace in Sydenham said its height was 363 feet, the diameter 100 feet from the base was 15 feet, and the bark at the base was 18 inches thick. A little twisty live *Sequoia* now grew at the base of Mother-Sacrifice Tree. Fuzzy hazels and scarlet dogwoods screened the enormity. Nearby lay a fire-hollowed, supine log called "Father of the Forest" that felt, when I walked through it, like a cool and eerie lava tube. Both Mother and Father had seen better days. Two other prone sequoias that had once held fire between them now made another tunnel. Sap ran down hard and shiny, looking like liquid obsidian.

Fire plays an important and largely beneficial role in the lives of the Big Trees. During the two days a week I spent on information duty at the General Sherman Tree, the largest sequoia of all, the most frequently asked question was, "What are all the black marks on the trees?" The answer, of course, was fire. *Sequoiadendron gigantea* is especially fire-adapted, in fact requires fire for successful germination, and the bark is therefore flame-resistant. I once crammed ninety people inside the hollow bole of one burned-out mammoth on a ranger walk. Another nearby tree had 90 percent of its mass burned away, yet lived. The first controlled burns to encourage sequoia regeneration took place in the national park that summer. Now they are an accepted part of conservation management. In the Great Plains, however, zealous prairie-burning sometimes threatens rare grassland butterflies, whose fire resistance is less than that of many plants, and fire represents one of the greatest dangers to winter-fat monarch clusters, fountains of tallow awaiting ignition.

The monarch in the Calaveras visitor center collection was accompanied by a label saying that it occurs generally from May to October and that "a wanderer can show up anywhere." I kept that thought uppermost as I rolled deeper into the Sierra Nevada. Incense cedar and sugar pine gave way to white and Shasta fir as I climbed through 5,000 and 6,000 feet into the glitter of sugarstone granite, with fresh snow up above. Band-tailed pigeons burst out of Hell's Kitchen. At home they would be gathering now at the oak mast with the Steller's jays. Here they came for the acorns of a little prostrate oak. Tamarack, population 9, had lodgepole pines but not larch. Tamarack is another name for the deciduous conifer *Larix,* or larch. It is often confused with tamarisk *(Tamarix),* the salt cedar of my recent close acquaintance, and tamarind *(Tamarindus),* an East Indian legume whose pod makes a bracing drink.

At Alpine Lake, all development well behind, the road went suddenly narrow, with no center line, and crossed the Tahoe-Yosemite Trail. "Ebbets Pass Ahead / Very steep, narrow, winding road / Vehicles over 25′ long not advisable." Mosquito Lake was ringed by white boulders, an elfin ranger station among them, whose occupant would no doubt pay for the scenic assignment in blood. Once over the Pacific Grade Summit, 8,050 feet, the road fell away in a thousand-foot roller-coaster ride down from Pacific Valley to Hermit Valley and the North Fork of the Mokelumne River, and over what surveyor G. H. Goddard called "a route of great promise — probably the best one for a transcontinental railway." No railroad or emigrant trains ever came this way, but the Silver City stage used it to serve the mines in 1864. Great-trunked junipers and bristlecone pines thinned toward timberline at Ebbets Pass, between the Stanislaus and Carson watersheds. Fresh snow wrapped the high bare shoulders of the High Sierra in a landscape as spectacular as Squaw Valley, but not yet ruined like that sad place.

I stepped into the Mokelumne Wilderness Area on the Pacific Crest Trail, snowy in the shade, muddy in the sun. It wasn't hard to imagine monarchs here; it was a lot like the Sierra Chincua, and at similar elevation. But 9,000 feet in *this* sierra is equivalent climatically to around 13,000 feet in middle Mexico; monarchs might last long enough here to make suitable orange and black Halloween decorations in the trees, but the same kind of winter that downed the Donners would make quick work of them by Christmas.

Through green and gold aspens, the road dropped abruptly toward the basin. Kate Wolf sang "The Golden Rolling Hills of California" just as I left them and entered Nevada in the late-day sun, after miles of monarchless rabbitbrush along the Carson River. Redtails and kestrels peered from the telephone lines and poles. I'd heard reports of monarchs migrating through the Washoe Valley between Carson City and Reno. But it was late autumn here now, and even the common butterflies were gone.

I lit out from Reno as soon as I got there. Back on Carson City's margin, the Walmart and Wendy's, McDonald's and Merwyn's, only an hour from the Pacific Crest Trail, had made me want to scream. The rainbow welter of Reno's casinos was still worse. I regained righteous darkness in the territory of night by turning north toward Pyramid Lake. Ancient Lake Lahontan had covered a vast expanse of the western Great Basin (more than 8,600 square miles and as much as 900 feet deep) about 14,000 years ago. Pyramid is now its deepest remnant. I had imagined rimming its alkaline turquoise bowl in search of monarchs, but not by night. All I saw flying now was a poorwill, like a cross between a large moth and a small owl.

At midnight I rolled into Gerlach, "Gateway to the Black Rock Desert" — the well-known one this time, not the inky lavaland I'd crossed in the Sevier Basin of Utah. The motel was full for a planned jet-car assault on the land-speed record, but I had planned to camp anyway, so I headed out of town to the storied playa. Craig Breedlove's speed show compound glowed like a carnival, a luminous tent-and-trailer city. It held no attraction for me. I hadn't come to the place of which Gary Snyder has written, "Off nowhere, to be or not be, / all equal, far reaches, no bounds. / Sound swallowed away," for a lot of commotion. It was just my good timing.

Despite its name, the desert was pallid in the cloud-moon gleam. No fan of off-road vehicles, I felt uneasy driving across it, but I'd been assured that it was okay. Gary Snyder, Barry Lopez, and Stephen Trimble had all coasted to a stop out on the gritty plain to see what it was like, and the Cream of Wheat surface was crisscrossed as far as I could see with tire tracks. Powdermilk, shaking in the wind, blended right in. The night was cold. From where I parked, I could see just one light up on the slope of a mountain to the northeast. The dim light of my overhead provided the only answering pole in the night.

I read a few pages of John Janovy's travels with yellowlegs, the memorable part where he has followed the banded bird all the way to the Texas coast and watches it fly out over the Gulf of Mexico. He's come as far as he can, but still he follows on, plunging his new secondhand Mustang right into the surf. I once drove my future mother-in-law's new Chrysler into the Boulder River, but I knew I wouldn't go that far with Powdermilk — not after driving her to the moon after monarchs. Still, the Black Rock Desert at night felt like total immersion.

I was nearly asleep when I heard the rain. Still in my sleeping bag, afraid of getting really stuck, I drove to the edge of the ancient lakebed and bedded down again just off the slick lip of the dampening playa. I awoke at eight to lightning in the West and the North, where I'd soon be heading. After a walk on the sticky talc of the Black Rock Desert, my sandals weighed several pounds each, and I was considerably taller.

The speed freaks' caravansary, complete with bus, tractor-trailers, and vast motor homes, was slowly decamping too. The bus looked as if it might not get out for a while. The giant jetski had remained civilly stabled for my hours there, and now it hunkered into the mire as ravens beat against the wind above. In *Desert Notes,* Barry Lopez wrote of ravens, "At first light their bodies swelled and their eyes flashed purple." Now the light was dull. These ravens swelled all right, as their feathers parted and their beards ruffled on the air, but there was no flashing going on. It was difficult to get into any sort of Lopezian/Snyderesque hot-desert state of just being out there, in the wind and rain, with tracks everywhere, as on some Washington beaches where knobby tires outnumber sanderlings. Hard to be reverent with the would-be speedsters pulling out with would-be alacrity and some difficulty. I didn't give a spit about the speed record, but I would have loved to see the great beast try to go really fast in that greasy pudding.

The eponymous Black Rock loomed over the big, flat floor of Ancient Lake Lahontan, which had once wetted much of western Nevada. To say that it has since "dried out" is a relative statement; today, for example, the saturated putty could almost be called a lake again. Many species of butterflies have evolved washed-out looking subspecies in the alkaline desert, perhaps adapting to blend with their pallid surrounds. One of these, a pale variety of viceroy *(Limenitis archippus lahontani),* takes advantage of monarchs passing through to educate the local birds about the hazards of preying on large orange butterflies. A couple of big bushy trees grew way out on the edge of the desert, the only ones for miles. Monarchs *could* stop here, as easily as at Bonneville.

People come to the Black Rock Desert for many reasons. Its flatness beyond flat, its loneliness, its clean slate (until the tire tracks arrive), all attract an array of pilgrims. Whatever leads them there, some people are marked by the place. Stephen Trimble, in *The Sagebrush Ocean*, writes that a Reno car-wash manager told him he would never get rid of the smell of the Black Rock dust. I was marked by my visit, too, and I never have gotten the smell out of Powdermilk — but it wasn't the dust; it was garlic.

Leaving the desert, I came across a mountain of garlic by the side of the road, spilled bounty from a tipsy truck. The top layer was a little soggy and showed a green sprout or two, but thousands of cloves of perfectly good, dry garlic lay heaped beneath. For an hour I gleaned bushelsful, filling every spare bag and each empty cranny in my little car. I wished I had some gunnysacks for the roof rack. Walking in garlic, the good smell engulfed me much the way certain other layers had bathed my senses in these past weeks: lenses of water in the John Day River that took me warmer or colder as I rose through them . . . supple skins of aspen leaves clothing me in the Utah mountains . . . shifting sediments of living wings conducting their soft orogeny of monarchs all around me at Morro Bay. The sheets of scent sifted up, and as I drove away I was enveloped in the beneficent vapors of the Stinking Rose, growing stronger with each mile.

Gerlach boasted four casinos, no shops, the now-deserted motel, and a gas station. When I gased up, the excited attendant told me that Breedlove had hit 448 mph before the rain came, though the try for a record was aborted. I asked if they grew a lot of garlic at Gerlach. "Nah, they grow that shit at Orient," he said, with a dismissive wave.

Up canyons to high prairie cast in broad strips of olive, beige, and black. Buckhorn, Ravendale. Horned lark flocks, and one road-killed lark: those tiny devilhorns, for once up close. Wind, the smell of rain on weeds. Around a bend, beyond a pasture, another great white playa lay like pancake batter just starting to bubble. This one was fenced and hadn't a mark on it, and came with a rainbow above, a brilliant, full, one-eighty arching over grazing cows. That rainbow traveled with me for many miles, and I discovered that you cannot focus on a rainbow with binoculars. A magpie flashed black-and-white frames across the rainbow's colored scrim. I was moving north through Surprise Valley toward Oregon, looking for laggard monarchs on the wind. Had I seen the flicker of one, it would have been no bigger surprise than the intensity and longevity of that vividly banded ribbon of sky.

I passed back into California a little before Cedarville, where I spent my last eighty cents on a good cup of coffee. Northward took me to Fort Bidwell in heavy rain. Larry Flat, Lily Lake, Dismal Swamp, a handsome, boarded-up stone schoolhouse on a small Indian reservation. Farther on, in Oregon, lay Plush and Warner Valley and, beyond that, Frenchglen and the good milkweed-and-monarch country of Malheur. Having bypassed this area weeks back in favor of the Snake River route, I had hoped to see some of it now. But an Alaskan storm was blowing in, and I didn't want to be stuck east of the Cascades if the passes were blocked. Reluctantly, I veered northwesterly.

To cross the Warner Mountains, I started up a steep clay road, wet and slick in the rain. It was a fragrant passage through beautiful juniper and cliff country, and I watched a peregrine falcon stoop on a rock face full of rock doves. But it was just too damn slippery; I gingerly turned around on a steep greasy pitch with a deep ditch and a long dropoff. Momentarily, Powdermilk was mired in the gumbo grade, but the remarkable, quarter-million-mile, 1300cc motor pulled us out.

The next road over the mountains, to the south, crossed Fandango Pass. It looked a little more manageable, though its sign warned "Steep Grade" too. I flagged a very old Indian man who was driving down and asked him how it was up ahead. "Good," he murmured. Then, looking over Powdermilk, "Yah, you can make it!" and he gave me a toothless, warm smile. I took his word.

Fandango Creek swirled muddily, farther and farther below. The road was glassy on switchbacks, and I slithered and fishtailed, but by keeping up the momentum I was making it. Still, when I spotted a stand of showy milkweed, the first I'd seen in days, I braked to look it over without even considering whether I could get going again. Some of the leaves were green, but looked frost-nipped. I examined each one with care, and could scarcely believe my eyes when I saw a monarch chrysalis — the first pupa I had found myself on the whole long trip.

I badly wanted it to be alive. It was still green and gold, but not the spring-green of the one Thea had spotted in the Okanogan, its gold stars a little tarnished. The chrysalis of a butterfly is often likened to a sarcophagus — most inappropriately, since they are wonderfully alive inside. But finally I had to admit to myself that in this case the analogy was correct. The pupa was dead — frosted, almost certainly, just last night.

A few degrees warmer, a little more sunshine, and this straggler might have eclosed today. I could have seen a monarch flying in Surprise Valley after all. Whether it could have made it to any winter sanctuary this late was an open question. I remembered the night roost I had seen at Bear River Refuge nearly this late years ago, which had inspired my Salt Lake detour from the Snake on this trek. The latest leavers are at greater risk from early frosts and storms, just as the earliest returnees face the fickleness of spring.

It was late in the day. The wintry wind whipped the floppy yellow and green leaves. There would be no more monarchs. But this poor pupa, a little too late, linked the trip home to the trip out, the coast to the basin to the rest. A monarch from Fandango Creek, we presume, would almost surely migrate to the California coast. Yet from there it wasn't that far east to Elko and the Bonneville Salt Flats, from where monarchs steered me southeast — pointing, if not directly leading, all the way to Mexico. The ineluctable conclusion — that western monarchs migrate to Mexico as well as California and that their pattern is far more complex than we have long believed — draws directly from the butterflies, in concert with the land. The secret of what each butterfly can do, and how it reads the landscape and chooses the way to go, lay imbedded in the quick-frozen tissues of this very chrysalis.

If the monarch of Fandango Creek would not join the others in the winter clusters, it taught me again that quality Robinson Jeffers described in "November Surf" as "the dignity of room, the value of rareness." The monarch masses that Jeffers knew in Carmel, that I saw at Morro Bay, are marvelous in their thousands, but they can never teach that. It is good to be reminded that every individual counts, this failed chrysalis as much as any.

Cresting the Warners at last at Fandango Pass, I met the Applegate & Lassen Cutoff Trail. The Applegate, an alternative southern route to the Willamette Valley, which left the Oregon Trail east of Twin Falls, came to be known as "the damnable road." Though it avoided the perils of the Columbia Gorge, it cruelly directed settlers through almost impassable canyons and across the Black Rock Desert. I pointed my wagon down through "stately pines," as an Applegate pioneer put it, hung with chartreuse lichen. The valley below was autumn tinder, and the late sun emerged to ignite it.

The radio back on, I learned from the farm report that Bob Pyle, of Susanville, offered 105 breeding bulls at the Shasta Livestock Auction in Cottonwood. On my namesake life list I have a Robert L., a noted Hawaiian ornithologist; a Robert W., who wrote the flora of Cape Cod; a plain Robert, who was famous in roses; and once, in the Trobriand Islands of Papua New Guinea, in the smoke of a sing-sing campfire, I even met a Robert M., of Delaware, interested in birds and butterflies. But this was my first bull-breeding Bob.

Cumulus fields rose above the sun, a bright shiny cap to the cloud wall predicting heavy weather. By the inland sea of Goose Lake stood Willow Ranch, with tumbleweeds blowing, longhorns in winter coats, and a fence hung with old pots and pans. Then came New Pine Creek, the other end of the only road that had turned us back. I struck Oregon in rain, no surprise there, which quickly became sleet. In Lakeview I inquired at a gas station about the best route west. The attendant, in an insulated jump suit, suggested the Ashland route. "Good road," he said. "Snow level at five thousand feet."

"But it's five thousand feet right here," I said. "There's slush south of town."

"Hail, here," he said. "Big splats. She's a comin' all right. An' we'll be looking at 'er a long time. Yep, comin' too early fer me."

I could head north to Bend, then Hood River, and on down the Columbia, and never have to cross a high pass. But I hated to get stuck east of the mountains without chains, and if a bad storm came in, that route could be a harsh one. The Columbia Gorge might even be closed. It seemed best to get across the Cascades here, before the weather got any worse. So I turned west, into the wind.

Soon I was driving in horizontal snow and high wind. No one else seemed to be going this way. I stopped to call Thea at Quartz Creek House, 5,504 feet, where the sticky snow was stacking up. Farther on I slowed for a cat in the road; from its posture I thought it was hurt, but it was just eating a mouse, and ran off with it. Lucky cat, because a school bus was coming and I couldn't have swerved. Then a huge brindle cow leapt out in front of me and I missed it by inches, slamming on the brakes and skidding. It would have been garlic beef. The traveling museum of natural history was in great disarray, the dancing girls from the Sanpoil River all awry on the dashboard.

Outside Klamath Falls a policewoman stopped me about the headlight that had been out for weeks. She said at least three cars were in the ditch already, and that was on the *better* of two possible passes. She recommended I take the road to Medford. "But really," she said, "I wish you'd just get a room for tonight."

Very few cars or trucks were coming over the Medford road, and several were indeed in the ditch. It was hard to maintain traction without snow tires or chains. The driven flakes were mesmerizing with my brights on (which worked), and the one low beam was too dim to see by. It was like driving into the tail of a comet. I kept on, now and then passing a creeping glacier of a semi slip-sliding the other way. This was the worst blizzard I'd ever been dumb enough to try to drive in. Through the whiteout, I began to see things that weren't there in the universe of snow. At one point the flakes became butterflies . . . more, even, than all the monarchs there ever were. I was afraid of dropping off, narcotized by the blinding wall, but certain that I shouldn't stop.

At last I was over the shoulder of Brown Mountain and the Cascade Crest. But I still had ten miles of 5 percent downgrade boring into the hypnotical snowburst. The trucks crept slower still. The blizzard spit me out at the western end of the Lake of the Woods Highway at 1 A.M., and an hour later I camped for the night at Rogue River — not in one of the state park's cozy, heated yurts but in garlicky Powdermilk, for the final time of the trip.

In the morning I was so weary from the blizzard that I slept until ten. The chrysalis had gone dark olive, almost lost its green, and the gold was going fast. In the cold, heavy rain, the only thing gold was the embarrassed tarnish of oaks in the Siskiyous. There are monarchs in this southwest Oregon country, though not at this season. In her *Spirit of the Siskiyous: The Journals of a Mountain Naturalist,* Mary Paetzel writes that in June "I visited the milkweed patch by the river, and the monarchs are back." But in late October "a lone monarch or two sails forlornly over the milkweeds, wings ragged, color faded, flight slow and hesitant. These are the old ones, the remnant . . . after a night of silver frost, the wilted leaves of the plants will lie in golden mounds beneath their bare stalks, and somewhere not too far away, the tattered wings of the monarchs will rest close to the frozen earth, as lifeless as the faded yellow leaves."

I stopped at a Grants Pass café, forgetting I was broke. The coffee smelled wonderful. A well-dressed woman from Ashland was placing an order beneath the "Espresso" sign at the counter. "Give me a triple tall skinny latte, please," she said. "Decaf."

The veteran waitress who took her order disappeared for a few minutes. Then she came back and asked the woman, "How do you want that, honey? Like a straight cappuccino?"

"No!" said the woman. "I ordered a latte, with —"

"Never mind, honey," said the waitress. "I make them all the same."

By the time the waitress got to me, I realized I was out of money and turned to go. Then I remembered the aluminum I had salvaged from Gloyd Seeps and the fifty-eight cents I had squirreled in the ashtray. Coffee was fifty-five cents, and there is no sales tax in Oregon.

I stopped at the old Portland-Sacramento stagecoach stop, the Wolf Creek Tavern, to call Thea and let her know I was safe. Rolling north, I passed Riddle. I passed Drain. I passed Roseburg's drive-in theater and wildlife safari, endangered phenomena I'd seen maybe three each of on the whole trip. Mostly I-5 was a big smear, a watercolor wash of sodden red and yellow maples, wet maroon madrona bark, and caramel oaks. Detouring on old U.S. 99 took me through Cottage Grove, a town whose pretty name actually matches its appearance. It was all white porches full of jack-o'-lanterns and, in the historic district, covered-bridge banners in bright pink, blue, and green. But the bucolic quiet was broken by a particularly awkward noise arising from Powdermilk's engine. Mike at Cottage Grove Auto Service told me her water pump was dying a horrible death. He sent me to Ernie at Dan's Automotive, who said it would be about an hour. While Ernie performed the transplant, I took a walk around town.

It was good to be back in the Northwest damp after months of aridity. Good too was the deep green moss in the sidewalk cracks, the rich red of the cherry leaves, the tiny-leaved ivy spilling over an old stump, the smile from a woman going into a shop. In the W. A. Woodard Memorial Library, I found my favorite monarch book, Jo Brewer's *Wings in the Meadow.* I settled into a child's chair with the book on my lap and read of "the wings of flame, rising to the sun," and how "the children of Danaus and his whole tribe flew along the beaches of New England and crossed small bodies of water with the wind. When they reached the Cape Cod Canal, they flew along its banks. . . . The group was not always the same, for some diverged to other routes, and found other Monarchs along the way, and it was never possible to know where the Monarchs along the endless stretches of sand and in the nearby countryside had begun their journey." *Amen.*

Back at Dan's, the car was almost ready. When I thought of all the deserts and mountain passes where the head gasket or the timing belt could have blown, a shot water pump in Cottage Grove with a library at hand seemed almost a benison. Soon I was back on the interstate for less than $150 on the gas card. The sun came out on scintillant green fields in the Willamette Valley, scrubbing the lambs white, setting off red barns against the steely Cascades. The light was pure Turner over the Coast Range. This was as far north as monarchs are known to breed commonly near the coast, but they were long gone now. Rain fell through the sun at Scio, and against brown clouds a *triple* rainbow appeared.

What looked like a fire on the left was a backlit blaze of vine maples through the low sun. But then there really was a car on fire, spewing acrid smoke across the road, then a five-car wreck in the southbound, and beyond that a two-car, blocking. It all looked undeadly, but traffic was backed for many miles, all the way through Salem. I was glad to be going the other way. I swung past Portland like a rocket borrowing the gravity of a handy planet. Entering the Columbia River's lower gorge, westward out of Longview, I breathed clean fog.

And at nine, greeted by stars, a nearly full moon, and Thea's warm embrace, I was home.

Recovery

After weeks of rain, the last days of October came to western Washington on a week of solid sunshine. Good for spreading the garlic out to dry, for unpacking, for coming home. Cleaning out Powdermilk, I checked the odometer. I had driven her a grand total of 8,647 miles. Adding in the 815 miles we'd put on Thea's truck meant that I had made a migration of 9,462 miles in fifty-seven days. Thea came outside to see what I'd brought back. She was thrilled with the garlic, selectively interested in the rest of it. I proceeded to list the contents, extracted mostly from the dashboard.

Inventory:
Traveling Museum of Natural History

Coyote melon, cotton boll, creosote sprigs, bristlecone pine cones (open and closed), giant sequoia cones (green, brown, chickaree-gnawed), Arizona sycamore bark, California sycamore bark and leaves, eucalyptus bark, eucalyptus leaves (red, tan, green), eucalyptus cones, pink dogwood leaves, screwbean mesquite clusters, four juniper berries, prostrate oak acorn-cap.

Tootsie Pop sticks, several meaningful twigs, toothpicks, used and not, wings of road-killed queens and checkered white, uncracked pistachio, California buckeye seed, showy milkweed leaves and exploding seed pods, faded leaves (many) of aspens and four species of cottonwoods (black, willow-leaved, Frémont's, Sargent's).

Boat-rope heart, pink duckbill salmon lure, pink sword swizzle

> sticks, one brass earring, three marbles (blue-green with cat's eye from hermit's village, yellow-green cat's eye, purple-green-orange-white with conchoidal fracture), one copper Levi's button, two shards of Talavera pottery, one bronze toad, several hair ties.
>
> Feathers of Steller's jay, black-billed magpie, raven, red-shafted flicker, cardinal, poorwill, dove, gull, and chartreuse boa, one raccoon's tail, one gold and olive-black waterbug shell with extremely sharp ventral thoracic spur, one limpet, three species of river clams (Columbia, John Day, Colorado).
>
> Three dancing girls, two monarch stickers, two monarch magnets, four other monarch effigies, and lots of rocks: coral pink sandstone, pink feldspar crystal, salt crystal from Bonneville, quartz crystal, agate, mica, Black Rock Desert #1 pebbles, Black Rock Desert #2 pebbles; three pennies.

Thea surveyed the lot. Then she said, "And those things all get kept?"

That evening Thea asked me what I thought I had learned in these two months spent in the company of monarchs. I said I'd gained a clearer sense of how these creatures actually live their remarkable lives, as caterpillars, chrysalides, and adults — feeding, molting, moving, surviving, and, especially, always adapting. The monarch was a real, functioning animal to me now, much more than the icon or symbol such a celebrated animal can easily become. I had lived with monarchs; they had been my guides.

I told her how I had reconfirmed the close relationship between butterflies and watercourses; they are not bound to the rivers but use them when it suits their purpose. And, although these particular butterflies do have a sun compass, they are definitely not drudges to a southwesterly orientation. It was quite clear that monarchs follow a spectrum of directions in pursuit of the southland, including southeast most certainly.

"And above all," I said, "I've satisfied myself that *not* all the monarchs this side of the Continental Divide migrate to the California coast; *some* fly south into Mexico. So the old model of the Rocky

Mountains as a kind of Berlin Wall for butterflies is bankrupt. The new model will be more complex, and a lot more interesting."

"Anything else?" asked Thea.

"Oh, yeah: let sleeping bridges lie, approach the Border Patrol with care, and stay the hell out of late-night blizzards."

"So none of your tags have been recovered?" she said.

"Chances were always small," I said. "And it might still happen."

One year later to the day, on October 26, 1997, thirteen-year-old Jeremy Lovenfosse and his mother, Natalie Helms, saw two monarchs fluttering over a road in Sea Cliff, south of Santa Cruz, where beautiful homes and eucalyptus trees rim the sea cliffs above the Pacific Ocean on Monterey Bay. One of the monarchs flew off, but the other remained in the street, looking somewhat damaged, as if hit by a car.

Jeremy picked up the small monarch, tucked her in his bicycle helmet, and took her home. He placed her outside on a clump of flowers in hopes that they might give her the nourishment she needed to fly off and join the other monarchs roosting nearby. A couple of days later, when he found the monarch dead on the ground, he read the tag on her left forewing and called the 1-800 number of the Monarch Program to report that he had found tag no. 09727.

A month before, on the cool, gray morning of September 26, David Branch and I had driven east up the Washington side of the Columbia Gorge to tag monarchs. I was using a new kind of tag from the Monarch Program with a 1-800 number instead of an address, and hoped to use up my stock before the end of the season. At a windsurfing site known as Doug's Beach, I nabbed a fresh big female monarch nectaring on a bright clump of purple asters. She received tag no. 09726. David's dog, Karma, rolled in a dead salmon and was summarily banished to the rear of the Jeep.

We arrived at North Roosevelt Petroglyph Park — the fish patch — at 3 P.M. The afternoon was cool, in the low sixties, mostly cloudy, with a stiff west wind. Down in the burdock/milkweed swale, David and I double-teamed a monarch on the wind. It perched twice in branches of a dead shrub before David finally caught it on the wing. I tagged the very fresh little female and released her on goldenrod. She flickered into the strong breeze and vanished behind a wall of Russian olives, branded forever with the number 09727. When she left Klickitat County, she was 690 air miles from Santa Cruz.

Butterflies do not fly in straight lines, at least not for long, so the actual distance she flew could well have been much more. Nor can we know when she reached her terminus. If you assumed a straight line and arrival the morning Jeremy found her, you would get twenty-three miles per day; but she had probably already been there for a couple of weeks before she was found, since monarchs are thought commonly to average forty to one hundred miles per day. Dr. Urquhart, for example, tagged a butterfly in Ontario in 1957 that was found in Atlanta, 740 air miles away, eighteen days later.

Taking a reasonably direct route, she would have crossed the Columbia River, passed down the John Day Valley to near the Painted Hills, then flown across the Ochoco Mountains and into the vast lake-and-desert region of south-central Oregon east of Newbury National Volcanic Monument and Crater Lake National Park. If the westerlies took her, she would have passed over a corner of the Cascades — where I found the blizzard — between Lower Klamath and Goose Lakes along the border; if she stuck with lower elevation and shorelines, she'd have sailed very near Fandango.

Once in California, #09727 had to cross the southern Cascades or the northern Sierra Nevada. She might have remained on the east side past Pyramid Lake to Tahoe and then up Donner Pass, a route monarchs are known to use, but more likely she made the passage between Mt. Shasta and Mt. Lassen, where the range breaks down. Of course she may have crossed at great height, well above the summits, rising on morning thermals, gliding down into the lowlands as the air became too cold.

However she crossed the mountains, our migrant now found herself in the Central Valley. There she traded supercooled mountain air and the threat of storms for a new set of challenges: agricultural chemicals, trucks and automobiles, and all the other hazards of the great breadbasket. To reach the sea safely she had to cross between eight and twelve freeways and the habitations of millions of people. And again, she may have performed this on high, oblivious, descending only when the weather, the night, or her need to feed dictated that she do so.

Finally, she would have sensed the onshore breeze, perhaps crossing the shoreline and returning to land. Once in the embrace of that lovely ocean crescent that has meant shelter to monarchs for at least 150 winters and perhaps many more, #09727 may have perceived the massed color, the movement, or the attraction pheromones of her own kind, and joined them in one of several possible winter roosts near Aptos. Her journey complete, she settled in for the fall and winter.

In the early maritime spring, the Roosevelt monarch would have mated and headed back out in search of milkweed to lay her eggs on, continuing as far as she could. But for her, spring would not come. Having survived thousands of risks to make it here, #09727 was hurt in what should have been her safe haven — probably struck by a vehicle as she sailed out on October 26 to seek nectar or moisture or simply to fly on the sunny breeze. Her companion, the one that flew off when Jeremy spotted her, may have been a male eager to mate months earlier than most; perhaps it was his aggression that drove her into the path of a car. Or maybe he spotted her after she was struck, fluttering in the road, as Jeremy did. If human care could have saved her, she would be there yet, one among thousands in a monarch-bedecked eucalyptus.

The specimen is now in possession of the Monarch Program. As it turns out, #09727 is the first instance ever of a monarch tagged in Washington State and recovered in California. It is also the first recovery of a *nontransferred* monarch — that is, one that originated at the point of first capture or got there under its own power — from the Pacific Northwest, other than Faye Sutherland's and Mary Henshall's monarchs from Boise. This butterfly's great significance was its natural, point-to-point flight. And more: while the Boise monarchs were reared from eggs and larvae found on milkweeds, #09727 was netted.

Our fish-patch female was the first wild adult tagged outside California and recovered within California in the known history of monarch tagging.

Number 09727 shows that some Northwest monarchs migrate to, and presumably from, the West Coast, as we have long suspected. Some other western monarchs appear (through successive individual vectors) to fly from northwest to southeast and at least some enter Mexico. My own observations were later confirmed by Rob and Eve Gill, experienced butterfly enthusiasts from Prescott, Arizona. They brought to my attention a memoir entitled *Oh Beautiful, Cruel Country,* by Eva Antonia Wilbur-Cruce, a pioneer near Arivaca, Arizona. She describes, near the Mexican border, near the turn of the last century, what she thought was "a golden brown fungus" on trees that, when touched, became "a brown cloud of monarch butterflies" that lifted up and "made a golden lace pattern in the sky."

Almost a hundred years later, and two years after I was there, the Gills themselves saw monarchs "flying very purposefully" south in Organ Pipe National Monument. Four of these were adjacent to the border between Quitobaquito Springs and Gachada Line Camp. And on October 5, 1998, Rob and Eve observed what I had been watching for in the same place — a monarch crossing the United States–Mexico border heading south, 2.7 miles west of the Lukeville/Sonoita Port of Entry.

David Marriott of the Monarch Program suspects that these butterfles may end up in Baja rather than mainland Mexico and may then cross to the coast and fly north to California in springtime. But for all we know, they may continue southeast to the Transvolcanic Ranges or to some winter site as yet undiscovered. Wherever their destination, I believe we will find eventually that Mexican monarchs returning in the spring enter the Southwest, some of them carry on to the Great Basin and Snake River Plain, some perhaps even invading the inland Northwest. Initial tests by Andrew Brower have failed to show significant DNA differences between eastern and western monarchs, and, as my findings show, they are hardly the monolithic, never-the-twain-shall-meet entities they were long thought to be. Yet the ways in which they mix are far from well understood, and they may yet be found to have biological differences.

I cannot say how much of the western population originates in Mexico rather than California in the spring or migrates to Mexico in the fall. Maybe most western monarchs *do* go to and from the West Coast. But in the past, conclusions of this sort have been based on assumption and repetition, at worst, and largely unreliable transfer recoveries, at best. At least now we have concrete evidence — one solid record — to show that such flights do occur. Perhaps someday there will be a recovery in Mexico from Okanogan or Maryhill or Farewell Bend. Only then will we know for sure that some Northwest monarchs share in a biological odyssey of a scale rivaling those of the salmon, the eel, and the crane.

Much more work will be necessary, in the lab, in the field, and in the mechanisms of the monarchs themselves, before we can fill in the picture. New tools will be brought to bear. Radar techniques have been perfected recently for detecting monarch flights at great heights and in clouds. Isotopes isolated from the wings of overwintering monarchs in Mexico show that the majority originated from midwestern areas with similar carbon and deuterium profiles in the rainwater; such studies could be replicated for the West. Further DNA analyses might finally support a genetic basis for destination, and additional magnetic investigations may show the degree to which monarchs' choices are directed by the earth itself. But the simple combination of the net and the sticky tag, primitive but sound, will continue to provide the foundation.

I may or may not go following monarchs again, but I know that Marsha and I will continue catching and tagging monarchs, hoping for more recoveries. In the meantime, the flight of this one butterfly of passage on her "wings of flame, rising to the sun," this vision of Danae touched by the golden rain, has made my journey whole.

AFTERWORD

How lucky I was, to go afield throughout the autumn of 1996, chasing monarchs. As you will read below, few of the falls since then would have worked as well, and several would have been downright fruitless. Even in 2008, when I undertook an even bigger, longer butterfly caper—afield all year long to pursue the first-ever butterfly Big Year (*Mariposa Road,* Yale University Press) and when monarchs turned out to be the most frequently sighted species of all (92 days out of 365)—relatively few of the many seen were in the West.

But that comes later. What is most important to note for the happy occasion of the republication of *Chasing Monarchs* by Yale University Press is this: its conclusions have only been confirmed, indeed reinforced, by findings since. All you have to do is check out the current migration map of Monarch Watch, and note the arrows from the West toward Mexico, to see that this is so. When these arrows first appeared, it was a great satisfaction to me. The monarch chase had overturned a long-held tenet of American natural history—that of the bicameral, "Berlin Wall" model of monarch migration that said all eastern monarchs go to Mexico, and all those born west of the Rockies, to California—and helped replace it with a subtler, suppler model showing not a complete blending of the two but a fundamental connection between them through Arizona—and why should we ever be surprised by greater connectedness? There are, after all, few true dualities in nature.

The Monarch Watch map tells the tale, and some other monarch maps (such as that of the vital young program Monarch Joint Venture) are beginning to get it right, too, though you still see the erroneous old maps trotted out even yet. That ancient, oft-repeated, and incorrect canon was a classic *factoid,* defined by the *Compact Oxford English Dictionary* as "an item of unreliable information that is repeated so often that it becomes accepted as fact." The big monarch chase challenged that particular factoid not by doing any great science but by going out of doors, among the monarchs, with open eyes and mind.

And yet, there was some science involved. I spared my readers from the numbers and data generated by the chase, but these have appeared in two peer-reviewed papers. The first, "Interchange of migratory monarchs between Mexico and the western United States, and the importance of floral corridors to the fall and spring migrations," was coauthored with Lincoln Brower. It appears in the book *Conserving Migratory Pollinators and Nectar Corridors in Western North America,* edited by Gary Nabhan. This paper gives the data and conclusions from *Chasing Monarchs* and proposes a new model for the western migration. Brower also introduces the rather shocking but more and more plausible hypothesis that the Californian clusters may be a metaphenomenon that flickers on and off, depending on occasional refreshment by Mexican immigrants shifted westward by prevailing winds, as eastern warblers were in 1996.

The second paper arising from *Chasing Monarchs* is published this year in *Monarchs in a Changing World: Biology and Conservation of an Iconic Insect,* edited by Karen Oberhauser and Kelly Nail (Oxford University Press). The book is based on the proceedings of the Sixth Conference on Monarch Biology and Conservation (MonCon VI), convened in Minnesota in June 2012. My contribution, "Distribution and Movement of Monarchs in the Pacific Northwest," again draws on the findings of the 1996 monarch chase to place our Cascadian monarchs in the larger context of the West as a whole. It enables me to show and discuss my fundamental findings—the monarchs' vanishing bearings—in graphic form, which tells the tale dramatically. Here are nine thousand miles of adventure expressed in one neat graph (see figure).

So the conclusion asserted in the first edition of this book has been upheld by the data recovered, once analyzed: some (and perhaps

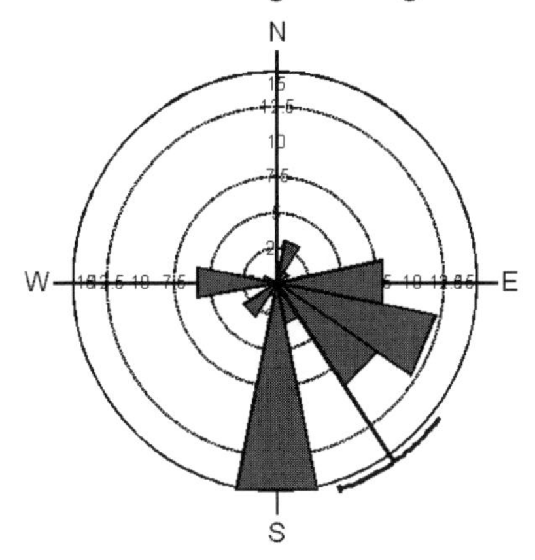

Vanishing bearings for autumn emigrants I observed in 1996 in the interior West, along a rough transect from Cawston, B.C., to Douglas, Ariz. The greater part of the measured movement was oriented S-SE. Data from *Chasing Monarchs* exercise (1999). Dates and capture localities appear in Brower and Pyle, "Interchange of Migratory Monarchs. . . ," 2004. Radii represent vanishing bearings of monarchs tagged and released in the West during the autumn of 1996, with the total number of individuals flying in each direction indicated by the length of each radius: total N = 62. Many more individuals were sighted; these data include only individuals with a discrete vanishing bearing as opposed to non-directional flight, surface foraging, or roosting. Note: the one WNW datum was for a monarch in central Arizona that was caught up in a mass movement of snout butterflies (*Libythea carinenta*).

many) western monarchs fly south or southeast toward Mexico, not solely west to California. In fact, this finding has been strengthened: first by naturalists Rob and Eve Gill, who watched monarchs actually crossing the US–Mexico border at Organ Pipe National Monument in 1998; and by the efforts of the Southwest Monarch Study (SMS) group. Beginning in 2003 the SMS, directed first by Chris Kline and now by Gail Morris, has sponsored extensive tagging of Arizona monarchs by many volunteers. Their early results were confused by mixing farmed and transferred California monarchs with wild Arizona monarchs, but those data are now being separated, and the trend is toward tagging wild monarchs only, as Gail Morris has told me. SMS has had several recoveries of wild Arizonan monarchs at Mexican winter sites. This is the first evidence of wild monarchs from west of the Continental Divide, tagged and released on their home ground (that is, not farmed or transferred), ending up in the Mexican overwintering colonies. Wild Arizona monarchs have also migrated to California coastal colonies. All this is a far cry from Fred Urquhart's map of 1977 (and many other maps since), which showed Arizona as a big blank for monarchs.

And now there is a chance that we could see additional recoveries from the inland Northwest—even unto Mexico?—thanks to studies initiated by David James of Washington State University. James has conducted extensive studies of the recently migratory monarchs of Australia and published many papers on them. Now he works on biocontrol in Washington vineyards and is hoping to parlay this work into extensive butterfly-friendly habitat benefiting monarchs and other butterflies. Toward this goal, he has developed a program in conjunction with the state department of corrections, whereby inmates of the state penitentiary at Walla Walla rear monarchs en masse to bolster wild populations. The social rewards of the program have been obvious and dramatic. Beyond that, by tagging and tracking them, James hopes to learn more about migration routes, their movements, nectar corridors, roosts, and other habitat components. Along the way, he intends to begin a new tagging program for the region that will enable others to take part, in hopes of taking our understanding still further. Although the vanishingly small number of monarchs in Washington in recent seasons has forced him to rely initially on breeding stock from southern Cascadia, James is working to source his project locally as soon as possible.

And this brings me back around to how lucky I was to chase monarchs when I did. The year 1996 did not offer up large numbers of monarchs for me to follow, but there were *some*, and it seemed a monarch almost always popped up where and when I needed it most: crossing the Columbia River, materializing out of the Bonneville Salt flats, beating across the desert toward the border. But such has not been the case lately. Both the Thanksgiving Monarch Counts in California, coordinated and compiled by Mía Monroe, and the World Wildlife Fund—Mexican government counts in Michoacán have reported in recent years the lowest numbers of wintering monarchs ever recorded.

Why is this? Many reasons may apply, including development of overwintering areas and loss of milkweed in California. As for the Mexican overwinterers, GMO (genetically modified organism) crops in the Midwest and ongoing logging in Michoacán play major roles, as Lincoln Brower describes in the Foreword. There is some welcome evidence that large-scale illegal logging in the Mexican reserves may be diminishing at last. Meanwhile, the threat from at least three genetically modified crops—Bt corn and Roundup Ready soybeans and corn—is only getting worse. Bt corn has been modified to carry the gene of a bacterium (*Bacillus thuringiensis*) that produces a toxin lethal to the larvae of Lepidoptera. The argument has been made and tested that toxic corn pollen reaches milkweed on the wind, killing monarch larvae. The loss to monarch numbers from this source is uncertain, but the injury from Roundup Ready soybean and corn is incontestable. By rendering soybeans and corn plants resistant to powerful herbicides, the new genes enable (and virtually require) greatly increased application of such chemicals across the Midwest soy and corn belt. This is how transgenic crops have made the heartland inhospitable for monarchs: the very region known to be the breadbasket for the species in North America. Because all of these GMO crops and the herbicides they bring on are products of one company, the name Monsanto has become the antithesis of monarchs in many people's minds.

Everything I have written about monarch conservation earlier still applies. But now comes a new threat, not as obvious in 1999 when this book was first published: global warming. The monarchs go to high elevations in Mexico or the Californian coast in late fall not to be warm but to be cool: to preserve their DNA and stored fat over the winter months. As the forests of the Trans-Mexican Neovolcanic Belt of south-central Mexico warm and dry, the winter clusters must rise higher and higher in elevation to find the cool, moist conditions they require. Some scientists project that the only Mexican sites capable of retaining Oyamel fir beyond 2050 might be the high volcanoes near Mexico City. But if cooked off the top of their current winter headquarters, would the displaced monarchs be able to find their way to Popocatepetl?

In the American West and Midwest, the great warming and drying are under way. It seems likely that continuing drought has much to do with the low numbers of monarchs in recent years. Both the absence of sufficient moisture for milkweed sprouting and succulence and the arid inhospitability of the migratory flyways stand to stem the natural and rapid replenishment of which insects are capable. Climate change is implicated as well in extreme weather events that can cause major mortality, such as hurricanes during migration. Great winter storms in Mexico have left deep drifts of dead monarchs, up to 80 percent of the population, in some years. And yet, as Barbara Kingsolver has posited in her novel *Flight Behavior,* perhaps monarchs will respond to warming by shifting their locus of winter activity. This could work either fer 'em or agin 'em, depending on the reliability of conditions in any new winter resort they could find. In the West, we could conceivably see clusters shifting north into the redwoods, or beyond.

All these factors make North American migratory monarchs an endangered phenomenon: a new concept when Lincoln Brower and I developed it in parallel around 1980. Their future is more uncertain than ever. Yet along with new and bigger threats have come novel and expanded efforts for their conservation. The various projects of Monarch Watch and the Monarch Joint Venture, for example, are encouraging. I am particularly heartened by the native milkweed cultivation, propagation, and distribution programs initiated by the Xerces Society.

Supposing all the good intentions and labor of conservationists, in concert with the supple adaptability of the organism itself, can give them a future after all, I wonder whether anyone will ever again attempt to follow the monarchs? I envy whomever it might be. I'll never forget those crisp autumn mornings, awaking in a campsite or highway rest area or beneath a monarch tree, and wondering what lay ahead for me, somewhere down the road, that day. Would there be any monarchs? If so, where would they be going? Because where they pointed, there I would follow. Nothing else to do, nothing at all, but chase the cinnamon sailors, seeking Danae's gold.

I'd go in a flash.

APPENDIX

FURTHER READING AND RESOURCES

ACKNOWLEDGMENTS

INDEX

APPENDIX

Conserving the Monarch of the Americas

During my last visit with Roger Tory Peterson, the conversation came around to butterflies, as it often did with Roger. Everyone knows of his elemental connection and contribution to bird study, but far fewer are aware of RTP's lifelong fascination and affection for butterflies. This he shared with his wife, Virginia, a fine butterfly gardener. We were discussing the decline of giant silk moths in the Northeast and the future of our butterflies. When we got to monarchs, Roger smiled. "The monarchs will take care of themselves," he said.

But will they?

A lot of energy is being spent on behalf of monarch conservation these days. In California, it is largely a matter of surveying and monitoring the wintering colonies, acquiring easements or other protections from development, and understanding and managing the dynamics of groves that make them acceptable to monarchs. In Mexico, initiatives in education, ecotourism, progressive forestry, and social justice, supported by solid research on monarch ecology, offer hopeful alternatives to the seemingly inexorable loss of monarch trees in the high forests where most of the North American population winters. Recently, a major meeting in Morelia, Michoacán, brought together many interested people, from local residents to biologists and bureaucrats. Secretary of the Interior Bruce Babbitt has announced his support for a trinational response to Morelia's challenge. But so far, action has been scarce, and the combined pressures of population, poverty, and profit represent major challenges to monarch survival. Lincoln Brower, our most distinguished monarch scientist, is by no means sanguine about the long-term security of the North American monarch migration.

Only twenty-four years have passed since Fred Urquhart's associates, Ken and Kathy Brugger, located the Mexican roosts, and little more than a century since the California clusters first became well known. The possibility that we could bring about the collapse of this astonishing phenomenon in such a short time makes me deeply dolorous. If we are to prevent such a bleak outcome, we need to know everything we can about monarchs, so we can take good care of every part of their migratory path. As Lincoln Brower recognized many years ago, there is no such thing as conserving monarchs by paying attention only to their winter havens; the entire migratory system must be understood, mapped, and cared for. The spraying and bulldozing of milkweed and nectar patches in the United States and Canada, both on the major breeding grounds and along the migratory corridors, may be just as dangerous as damage to the winter groves. Everything we can do to better understand and appreciate the movements of monarchs will help toward their conservation.

One recent development that hurts rather than helps is the transportation and release of butterflies—monarchs are the most desired and frequently used—for weddings, divorces, funerals, and other events. A lot of people have decided they can make money by catering to this unfortunate fad. Some observers have voiced concern over mixing locally influenced genetic stocks, and this is hypothetically worrisome if major DNA differences should be found among monarchs from different areas. Others express alarm at the prospect of spreading disease through transfer and release, and there seems to be a strong basis for this. But there can be no argument that transporting monarchs to places they never intended to go will give a distorted view of their natural range and movements. Butterfly atlas projects under way throughout North America furnish vital tools for conservation, since we cannot protect a feature unless we know where it is.

Only by tracking the natural whereabouts of monarchs can we know where they travel, how they make their way, and what they require. Yet every single monarch that hitchhikes with a wedding party threatens to warp our collective picture of the species. As tens of thousands of monarchs are released far from their birthplaces, the sum will amount to increasingly unreliable maps.

The Mixmaster approach to monarchs may damage their genetics, will probably spread serious diseases, and will certainly smear the clear view we need of their migration. For these reasons I strongly urge those who care about monarchs to avoid purchasing butterflies for release at events or in the classroom. Teachers can go out with students to find local caterpillars for rearing, as Faye Sutherland and Mary Henshall have done for decades in Boise; this gives students a genuine adventure that means a lot more than buying livestock from elsewhere and possibly doing harm in the process. And couples wishing to commemorate their wedding day responsibly should have their party toss rose petals, blow bubbles, or scatter the seeds of local milkweeds to the breeze. Butterflies, after all, are living organisms, not baubles or balloons to be used for pleasure and profit without consideration for their own requirements.

Through his long, observant life, Roger Peterson did not see monarchs decline as luna moths and regal fritillaries had. When he said that monarchs would take care of themselves, he was cannily recognizing the great resourcefulness and adaptability of these remarkable animals. Their host milkweeds, too, are resilient to many kinds of landscape changes. For these reasons, the monarch—established now in many parts of the world—will not become an endangered species in the usual sense. I would tend to agree with Peterson's prediction if conditions were to remain much the same as they have been. But the rapid spread of development, the proliferation of herbicides and other toxic chemicals, the introduction of transgenic herbicide-resistant crops in regions of heavy monarch breeding, Bt-corn whose pollen poisons caterpillars, and the continuation of logging in Michoacán all make me fear for their future.

It is my dearest hope that Roger Tory Peterson will have been right and that the extraordinary flight of the butterflies of passage will continue. But if you wish to help the monarchs take care of themselves, you can visit the Mexican monarchs and contribute liberally to the local economy. You can refuse to take part in transfers and releases, allowing monarchs to fly free where they will under their own power. You can participate in butterfly counts and monarch tagging and monitoring programs. You can join the organizations listed on pages 296–98. And, most important, you can help to nurture and protect the places where monarchs breed, travel, nectar, and roost. Only if enough of us care enough will the North American countryside remain replete with native monarchs and milkweeds.

FURTHER READING AND RESOURCES

The scientific and popular literature on monarchs is large. Here are some of the most significant titles, some I have referred to in this book, and some that have especially influenced me. Many additional titles are cited in the papers listed.

Ackery, P. R., and R. I. Vane-Wright. *Milkweed Butterflies: Their Cladistics and Biology.* Ithaca, N.Y.: Cornell University Press, 1984.

Baker, Robin R. *The Evolutionary Ecology of Animal Migration.* London: Hodder and Stoughton, 1978.

Brower, A. V. Z., and M. M. Jeansonne. "Geographical Populations and 'Subspecies' of New World Monarch Butterflies (Nymphalidae) Share a Recent Origin and Are Not Phylogenetically Distinct." *Annals of the Entomological Society of North America* 97 (2004): 519–23.

Brower, A. V. Z., and T. M. Boyce. "Mitochondrial DNA Variation in Monarch Butterflies." *Evolution* 45 (1991): 1281–86.

Brower, Lincoln P. "Ecological Chemistry." *Scientific American* 220, no. 2 (1969): 22–29.

———. "Monarch Migration." *Natural History* 86 (1977): 40–53.

———. "Understanding and Misunderstanding the Migration of the Monarch Butterfly (Nymphalidae) in North America: 1857–1995." *Journal of the Lepidopterists' Society* 49 (1995): 304–85.

———. *Para comprender la migración de la mariposa monarca.* Mexico City: Instituto Nacional de Ecología, 1999.

———. "Canary in the Cornfield: The Monarch and the Bt Corn Controversy." *Orion* 20 (2001): 32–41.

Brower, Lincoln P., et al. "On the Dangers of Interpopulational Transfers of Monarch Butterflies." *Bioscience* 45 (1995): 540–44.

Brower, Lincoln P., et al. "Decline of Monarch Butterflies Overwintering in Mexico: Is the Migratory Phenomenon at Risk?" *Insect Conservation and Diversity* 5, no. 2 (2012): 95–100.

Brower, Lincoln P., Mía Monroe, and Katrin Snow. *The Monarch Habitat Handbook: A California Landowner's Guide to Managing Monarch Butterfly Overwintering Habitat.* Portland, Ore.: The Xerces Society, 1996.

Brower, L. P., and R. M. Pyle. "The Interchange of Migratory Monarchs Between Mexico and the Western United States, and the Importance of Floral Corridors to the Fall and Spring Migrations." In G. P. Nabhan, ed., *Conserving Migratory Pollinators and Nectar Corridors in Western North America,* pp. 167–78. Tucson: University of Arizona Press and the Arizona-Sonora Desert Museum, 2004.

Calvert, W. H., L. E. Hedrick, and L. P. Brower. "Mortality of the Monarch Butterfly (*Danaus plexippus* L.): Avian Predation at Five Overwintering Sites in Mexico." *Science* 204 (1979): 847–51.

Calvert, W. H., and L. P. Brower. "The Importance of Forest Cover for the Survival of Overwintering Monarch Butterflies (*Danaus plexippus,* Danaidae)." *Journal of the Lepidopterists' Society* 35 (1981): 216–25.

Commission for Environmental Cooperation. *North American Monarch Conservation Plan.* Quebec: Commission for Environmental Cooperation, 2008. http://www.mlmp.org/Resources/pdf/5431_Monarch_en.pdf.

De la Maza Elvira, Roberto G. "La monarca del vuelo." *Ciencias* 37 (1995): 4–18.

Deming, Alison Hawthorne. *The Monarchs: A Poem Sequence.* Baton Rouge: Louisiana State University Press, 1998.

Dingle, H., M. P. Zalucki, W. A. Rochester, and T. Armijo-Prewitt. "Distribution of the Monarch Butterfly, *Danaus plexippus* (L.) (Lepidoptera; Nymphalidae), in Western North America." *Biological Journal of the Linnean Society* 85 (2005): 491–500.

Frey, Dennis, Kingston L. H. Leong, Eric Peffer, Robert K. Schmidt, and Karen Oberhauser. "Mate Pairing Patterns of Monarch Butterflies (*Danaus plexippus* L.) at a California Overwintering Site." *Journal of the Lepidopterists' Society* 52 (1998): 84–97.

Gibo, David L., and Jody A. McCurdy, "Evidence for Use of Water Ballast by Monarch Butterflies (*Danaus plexippus* [Nymphalidae])." *Journal of the Lepidopterists' Society* 47 (1993): 154–60.

Gottfried, Carlos. *Monarcas.* Mexico City: Condumex, 1984.

Grace, Eric S. *The World of the Monarch Butterfly.* San Francisco: Sierra Club, 1997.

Halpern, Sue. "A Fragile Kingdom." *Audubon* 100, no. 2 (1998): 36–44, 99–101.

———. *Four Wings and a Prayer: Caught in the Mystery of the Monarch Butterfly.* New York: Pantheon, 2001.

Heppner, J. B. *The Monarch:* Danaus plexippus *(Linnaeus).* Gainesville, Fla.: Scientific, 2013.

Hoth, Jurgen, et al. *1997 North American Conference on the Monarch Butterfly* (in English and Spanish). Montreal: Commission for Environmental Cooperation, 1999.

Jepsen, S., and S. H. Black. "Understanding and Conserving the Western North American Monarch Population." In K. O. Oberhauser and K. Nail, eds., *Monarchs in a Changing World: Biology and Conservation of an Iconic Insect.* New York: Oxford University Press, 2014.

Kingsolver, Barbara. *Flight Behavior.* New York: Harper, 2012.

Lane, John. "California's Monarch Butterfly Trees." *Pacific Discovery* 38 (1985): 13–15.

Malcolm, S. B. "Monarch Butterfly Migration in North America: Controversy and Conservation." *Trends in Ecology and Evolution* 2 (1987): 135–38.

———. "Mimicry: Status of a Classic Evolutionary Paradigm." *Trends in Ecology and Evolution* 5, no. 2 (1990): 57–62.

Malcolm, Stephen B., and Myron P. Zalucki, eds. *Biology and Con-*

servation of the Monarch Butterfly. Los Angeles: Natural History Museum of Los Angeles County, 1993. This book, which contains many important papers by various authors not otherwise listed here, is the product of the 1986 Second International Conference on the Monarch Butterfly, commonly referred to as MonCon II.

Merwin, W. S. "The Winter Palace." *Orion* 15, no. 1 (1996) 44–53.

Nagano, Chris, and John Lane. "A Survey of the Location of Monarch Butterfly *(Danaus plexippus)* (L.) Overwintering Roosts in the State of California, U.S.A.: First Year 1984/1985." Portland, Ore.: The Monarch Project (Xerces Society), 1985.

Oberhauser, Karen, and Kristen Kuda. *A Field Guide to Monarch Caterpillars.* Minneapolis: Department of Ecology, Evolution and Behavior, University of Minnesota, 1997.

Oberhauser, Karen S., and Michelle J. Solensky, eds. *The Monarch Butterfly: Biology and Conservation.* Ithaca, N.Y.: Cornell University Press, 2004. An important collection of papers from MonCon IV, the Monarch Population Dynamics Conference in Lawrence, Kans., in 2001.

Oberhauser, K. O., and K. Nail, eds. *Monarchs in a Changing World: Biology and Conservation of an Iconic Insect.* New York: Oxford University Press, 2014. The edited proceedings of MonCon VI, this volume will be the vade mecum of contemporary monarch science for years to come.

Perez, Sandra M., Orley R. Taylor, and Rudolf Jander. "A Sun Compass in Monarch Butterflies." *Nature* 387 (1997): 29.

Pleasants, J. M., and K. S. Oberhauser. "Milkweed Loss in Agricultural Fields Because of Herbicide Use: Effect on the Monarch Butterfly Population." *Insect Conservation and Diversity* 6, no. 2 (2012): 135–44.

Pyle, Robert Michael. "Migratory Monarchs: An Endangered Phenomenon." *The Nature Conservancy News* 33, no. 5 (1983): 20–24.

Pyle, Robert Michael, ed. *Symposium on the Biology and Conservation of Monarch Butterflies. Atala* 9 (1984): 1–45. This issue of the Xerces Society journal contains papers presented at the First

International Conference on the Monarch Butterfly (MonCon I), held in Cocoyoc, Morelos, Mexico, in August, 1981. It is published in both English and Spanish.

Pyle, Robert Michael. "The Historic Flight of Monarch #09727." *Monarch News* 8, no. 3 (1997): 1, 3–4.

———. "The Biogeography of Hope: Why Transporting Butterflies Is a Bad Idea." *Monarch News* 8, no. 6 (1998): 6–7.

———. "Under Their Own Steam: The Biogeographical Case Against Butterfly Releases." *News of the Lepidopterists' Society* 52, no. 2 (2010): 54–57.

———. "Distribution and Movement of Monarchs in the Pacific Northwest." In K. O. Oberhauser and K. Nail, eds., *Monarchs in a Changing World: Biology and Conservation of an Iconic Insect.* New York: Oxford University Press, 2014.

Rea, Ba, Karen Oberhauser, and Michael A. Quinn. *Milkweeds, Monarchs, and More: A Field Guide to the Invertebrate Community in the Milkweed Patch.* Union, W.Va.: Bas Relief, 2003.

Rendón, E., and G. Tavera-Alonso. *Monitoreo de la superficie forestal ocupada por las colonias de hibernación de la mariposa monarca en Diciembre de 2012,* pp. 1–6. Report: World Wildlife Fund–Mexico, Telcel, Comision Nacional de Areas Naturales Protegidas (CONANP) y Mariposa Monarca, Reserva de La Biosfera, Mexico City, 2013.

Riley, C. V. "Migratory Butterflies." *Scientific American* 38 (1878): 215.

Sakai, W. H., and W. H. Calvert. "Statewide Monarch Butterfly Management Plan for the State of California, U.S.A." California Department of Parks and Recreation, 1991.

Shepardson, Lucy. *The Butterfly Trees,* 1914. Rev. ed.: Monterey, Calif.: Herald Printers and Publishers, 1914.

Swengel, A. B. "Population Fluctuations of the Monarch (*Danaus plexippus*) in the 4th of July Butterfly Count, 1977–1994." *American Midland Naturalist* 134 (1994): 205–14.

Teale, Edwin Way. *Autumn Across America.* New York: Dodd, Mead, 1956.

Urquhart, Fred A. *The Monarch Butterfly.* Toronto: University of Toronto Press, 1960.

———. *The Monarch Butterfly: International Traveler.* Chicago: Nelson-Hall, 1987.

———. "Found at Last: The Monarchs' Winter Home." *National Geographic* 150 (1976): 160–73.

Urquhart, F. A., and N. R. Urquhart. "Overwintering Areas and Migratory Routes of the Monarch Butterfly (*Danaus p. plexippus,* Lepidoptera: Danaidae) in North America, with Special Reference to the Western Population." *Canadian Entomologist* 109 (1977): 1583–89.

U.S. Department of the Interior. *State and Federal Monarch Activities in the United States.* Washington: Department of the Interior, Policy and International Affairs, 1997.

Wells, S. M., R. M. Pyle, and N. M. Collins, eds. "Monarch Butterfly: Threatened Phenomenon, California Winter Roosts and Mexican Winter Roosts." *The IUCN Invertebrate Red Data Book,* pp. 463–70. Gland, Switzerland: International Union for Conservation of Nature and Natural Resources, 1983.

Wells, H., and P. H. Wells. "The Monarch Butterfly: A Review." *Bulletin of the Southern California Academy of Sciences* 91 (1992): 1–25.

Wenner, Adrian. "Monarch Butterfly Migration in North America: A Comprehensive Theory." Unpublished.

Williams, C. B. *Insect Migration.* London: Collins, New Naturalist Series, 1958.

Zahl, P. A. "Mystery of the Monarch Butterfly." *National Geographic* 123 (1963): 588–98.

FOR CHILDREN

This list represents a selection of the many available titles. Some of these are out of print but may be found in libraries or secondhand bookstores. These books are not solely for young readers; more people may have been charmed into a love of monarchs by Jo Brewer's *Wings in the Meadow* than by any other book.

Bach, Marcus. *I, Monty.* Honolulu: Island Heritage, 1978.
Brewer, Jo. *Wings in the Meadow.* Boston: Houghton Mifflin, 1967.
Calder, Joan Z. *Airplanes in the Garden: Monarch Butterflies Take Flight.* Santa Barbara, Calif.: Patio, 2011.
Catlin, Christine. *Raising Monarchs for Kids.* North Attleboro, Mass.: KidPub, 2010.
Frost, Helen, and Leonid Gore. *Monarch and Milkweed.* New York: Atheneum Books for Young Readers, 2008.
George, Jean Craighead. *The Moon of the Monarch Butterflies.* New York: Thomas Y. Crowell, 1968.
Gibbons, Gail. *Monarch Butterfly.* New York: Holiday House, 1991.
Harvey, Diane, and Bob Harvey. *Melody's Mystery.* Wilsonville, Ore.: Beautiful America Publishing, 1991. Text in Spanish and English.
Herberman, Ethan. *The Great Butterfly Hunt: The Mystery of the Migrating Monarchs.* New York: Simon and Schuster, 1990.
Himmelman, John. *A Monarch Butterfly's Life.* New York: Grolier, 1999.
Hoffman, Don. *Wanderer: The Monarch Butterfly.* Natural History Association of San Luis Obispo Coast, 1989.
Hutchins, Ross E. *The Travels of Monarch X.* Chicago: Rand McNally, 1966.
Lasky, Kathryn. *Monarchs.* San Diego: Harcourt Brace, 1993.
Lavies, Bianca. *Monarch Butterflies: Mysterious Travelers.* New York: Dutton Children's Books, 1992.
Lighthipe, Mindy. *Mother Monarch.* Gainesville, Fla.: Bugs, Beasts, and Botanicals, 2010.
Marsh, Laura. *Great Migrations: Butterflies.* Des Moines, Iowa: National Geographic Children's Books, 2010.
Ordish, George. *The Year of the Butterfly.* New York: Charles Scribner's Sons, 1975.
Ortiz Monasterio, Fernando, and Valentina Garza. *Mariposa Monarca: Vuelo de Papel.* Mexico City: Centro de Informacion y Desarollo de la Comunicacion y la Literatura Infantiles, 1984.
Pringle, Laurence P. *An Extraordinary Life: The Story of a Monarch Butterfly.* New York: Orchard Books (Scholastic), 1997.

Prior, R. W. N. *The Great Monarch Butterfly Chase.* New York: Bradbury Press, 1993.

Rea, Ba, and Carol Culler. *¡Monarca, Ven! Juego Conmigo.* Union, W.Va.: Bas Relief, 2011.

Romeu, Emma, and Fabricio Vanden Broeck. *Un bosque para la mariposa monarca.* Miami: Santillana Usa, 2001.

Rosenblatt, Lynn M. *Monarch Magic! Butterfly Activities and Nature Discoveries.* Nashville: Williamson, 1998.

Rotter, Charles. *Monarch Butterflies.* North Mankato, Minn.: Child's World, 1993.

Simon, Hilda. *Milkweed Butterflies: Monarchs, Models, and Mimics.* New York: Vanguard Press, 1969.

Weygant, Sister Noemi. *The Life Story of Little Maria and Her Offspring.* Duluth, Minn.: Priory Press, 1984.

ORGANIZATIONS

The following organizational, serial, and Internet resources are available to anyone wishing to learn more about monarchs, help conserve them, and follow developments in monarch science.

East Bay Regional Parks District (http://www.ebparks.org/). A source for monarch information, clusters to visit, and activities around the Bay Area. Search the website for "monarchs."

Journey North (http://www.learner.org/jnorth/monarch/). A connector for anyone wishing to follow the migration of monarchs and several other species via the internet. Many classrooms in Canada, the United States, and Mexico take part.

The Lepidopterists' Society (http://www.lepsoc.org). The primary organization for the study and exchange of information about butterfles and moths. Publishes *News of the Lepidopterists' Society* and *The Lepidopterists' Journal.*

Monarch Alert (http://monarchalert.calpoly.edu/). Project Monarch Alert is a citizen-based research project backed by graduate students and faculty from California Polytechnic State University.

They study the demography and population fluctuations of western monarchs through sampling of overwintering populations in San Luis Obispo and Monterey counties.

Monarch Joint Venture (http://www.monarchjointventure.org/). The MJV is a partnership of federal and state agencies, nongovernmental organizations, and academic programs that are working together to support and coordinate efforts to protect the monarch migration across the lower forty-eight United States.

Monarch Larva Monitoring Project (http://www.mlmp.org). The Monarch Larva Monitoring Project (MLMP) is a citizen science project involving volunteers from across the United States and Canada in monarch research. Developed and coordinated by researchers at the University of Minnesota, the project collects long-term data on monarch caterpillars and milkweed habitat to better understand how and why populations vary in time and space during the breeding season in North America.

Monarch Monitoring Project (http://www.monarchmonitoringproj ect.com). This very active arm of the Cape May Bird Observatory sponsors tagging of monarchs at Cape May, New Jersey, and tabulates and publishes their long-distance recoveries.

Monarch Program (http://www.monarchprogram.org). Based in San Diego, the Monarch Program monitors western monarch movements and overwintering sites, provides educational services, and publishes *Monarch News* monthly.

Monarch Watch (http://monarchwatch.org/). This site contains a mass of monarch information, chiefly concerning the East and Midwest, including the addresses of various other monarch groups. Monarch Watch tracks the migration, publishes a season summary, furnishes tagging materials, and runs DPlex-L, a monarch list-server.

North American Butterfly Association (http:/www.naba.org). Membership organization for butterfly watchers. Sponsors the Xerces/NABA Fourth of July Butterfly Count, and publishes *American Butterflies.*

Pacific Grove Museum of Natural History (http://www.pgmuseum

.org/). If you can't visit this "superb little museum," as the great nature writer Edwin Way Teale called it, visit the website to see what's going on with the famous monarchs in "Butterfly Town, U.S.A."

Southwest Monarch Study (http://www.swmonarchs.org/). Researchers in the Southwest Monarch Study are studying the migration patterns of monarchs in Arizona through extensive tagging and observation. They have already backed up the findings in this book by obtaining recoveries of tagged Arizona monarchs among the Mexican winter colonies.

Texas Monarch Watch (http://www.texasento.net/dplex.htm). Follows and reports on the Texas migration.

The Xerces Society (http://www.xerces.org/monarchs/). The international organization for invertebrate conservation. Beginning with the Monarch Project in the 1990s, the Xerces Society has supported California overwintering site surveys, habitat assessments, and habitat improvements, worked for easements and other protective measures, and published management guides and a review of laws that pertain to the protection of monarch habitat. In particular, Xerces has launched Project Milkweed, an initiative to enhance monarch breeding habitat by increasing the availability of native milkweed seed in regions of the monarch's breeding range where sources have previously been scarce. The website also contains the California Thanksgiving Count data, where to go to see monarchs, recent articles on monarchs, and much more information.

ACKNOWLEDGMENTS

To Charles Remington, Lincoln Brower, William Calvert, and John Lane, my teachers, for their great knowledge of the wanderer, steadfast friendship, support, and longstanding assistance with this and other monarch projects; and to Lincoln for his crucial reading of the text.

To Fred and Norah Urquhart, Edwin and Nellie Teale, Roger and Virginia Peterson, Jo Brewer and Dave Winter, Joan and Bill DeWind, and Floyd and June Preston, for the counsel of elders over many years.

To Walter Sakai, Dennis Frey, Kingston Leong, Karen Oberhauser, Orley Taylor, Mía Monroe, Barbara Deutsch, Julie Sidel, Christine Arnott, Don Davis, Susan Borkin, David Marriott, Christian Manion, Ray Stanford, Rodolfo Ogarrio, Fernando Ortiz Monasterio, Leonila Vázquez Garcia, Héctor Perez R., Roberto and Javier de la Maza, Mathew Tekulsky, Sue Wells, Marc Collins, Tom Lovejoy, Andrew Brower, Julian Donahue, Chris Nagano, Ro Vaccaro, Adrian Wenner, Sheri Moreau, Paul Cherubini, Peter Jump, Rick Bailowitz, Jim Brock, Robert Small, Mike Quinn, Steve Buchman, Perry Conway, Sue Halpern, and many others, for sharing their experience with and understanding of monarchs.

To John Hinchliff, Ann Potter and David Hayes, Jon Pelham, Jon Shepard, Dave McCorkle, Paul Hammond, Robin Cody, Dan Hilburn, Dan Carney, Vern Covlin, Patti Ensor, Paul and Vicki Runquist, Maurita Smyth, Sue and Jim Anderson, Dennis Strenge, John Coolidge, David and Jan Johnson, Bill Leonard, Chris Guppy, Claire Hagen Dole, Larry Everson, Ken Goeden, and the Northwest Lepidopterists Association for help on home ground.

To Helen Knight, Rob and Eve Gill, Karölis Bagdonas, Steve Kohler, Ron Hellstern, Steve Herman, Dan Pentilla, Steve Trimble, Ann Zwinger, Susan Zwinger, Terry Williams, Melody Allen, Katrin Snow, Susan Tweit, Mike Houck, Rick Brown and Ruth Robbins, Peter Steinhart, Brion Zion, David Myers, Mo Nielson, Wayne Wehling, Jolé Miller, Jeff Glassberg, and the United States Border Patrol in Douglas, Arizona, for tips, gifts, and valuable advice of many kinds.

To Martha and Mike Buckingham, Robert, Jane, Nick, Chris, and Claire Mennell, Ron Loiseau, Dennis St. John, Rosie and Cecil Carr, Barbara and Richard Kerb, Anne and Leon Martin, Jan O'Dell, Chris and Mike O'Brien, Caroline Wilson, Alison Deming, Roseann and Jonathan Hanson, Big Field village of the Tohono O'odham Reservation, Tamara Rarig, Susan and Ted Kafer, and the Jon Cristiansen family for their kind hospitality and assistance along the way.

To John Janovy, Jr., William Lishman, and Dan O'Brien, for their migration tales; Faye Sutherland and Mary Henshall for their valuable teaching and critical Idaho tagging records; David Branch and Karma, for going afield with me again and again; Gary Nabhan, for sharing monarch-cum-bat sightings and the stinking hot desert; Jeremy Lovenfosse and Natalie Helms, for their spectacular find of Monarch #09727; and Gail Morris and Chris Kline, for keeping me up on Arizona.

To all of my Orion Society and Forgotten Language Tour friends, too many by now to name, and to my writing group—Susan Holway, John Indermark, Pat Thomas, Jenelle Varila, and Lorne Wirkkala—for improving the book through their indulgent listening and perceptive comment, and for the communities of friendship and mutual sustenance they both give me.

To my ever-remarkable editor, Harry Foster, his astute colleague Peg Anderson, Katie Dillin, Anne Chalmers, and others at Houghton Mifflin Company, and my incomparable agent, Jennie MacDonald at Curtis Brown, Ltd., for imagining this book with me and seeing it through to eclosure.

To Jean Thomson Black, Samantha Ostrowski, Meredith Phillips, and their colleagues in New Haven for giving the book vibrant new life in the Yale University Press edition.

To Tom and Dory Hellyer, Fayette Krause and JoAnne Heron, Pattiann Rogers, Brian Doyle, Kim Stafford, Scott Sanders, Richard Nelson, Mark Garland, the Madduxes, Maxwells, and MRMs, Powdermilk, Marsha, and Danae, for fundamental and continual support of the chase.

To all those working to conserve monarchs; to the monarchs themselves.

And especially, for discovering the monarchs of Gallagher Flat, for her watchful and warm partnership on the first fortnight of the journey and for making the rest of it possible, for careful reading, close listening, Spanish translation, the lino-cut title-page motif, and everything else, to Thea Linnaea Pyle (1947–2013). I shall miss her beyond measure.

To all of you I offer my heartfelt thanks.

INDEX

A'al Waipia, 218–19
Abbey, Ed, 178, 185
Abbot, John, 147
Accidental migrations, 54–55, 96
Ackery, Phil, 50
Admirals, 232
  Lorquin's, 17, 124
  migration patterns of, 115
  mimicry of, 65, 193
  red, 60, 85, 112, 115, 172, 233, 252
  white, 20, 65
Adobe Pass, Nevada, 156, 158, 159
Aging/deterioration, of monarchs, 2–3, 98
Alar (wing) pockets, 80, 194
ALE (Arid Lands Ecology) unit, 76
Alfalfa butterflies, 122
Alfalfa fields, 130–31, 134, 149, 172
Allen, Melody Mackey, 246
Ambush bugs, 48, 60, 107
American Museum of Natural History, 202
Amphibians, populations of, 52
Ancient lakes, 163, 177, 259
Anderson, Andy, 15, 18
Anderson, Sue and Jim, 114
Androconial patches, 80, 86, 194
Anglewings
  satyr, 24–25, 118
  zephyr, 115, 120, 121, 174
Aphids, milkweed, 31, 36, 86
Aposematic insects, 32, 36
Arctic, Bean's, 18
Aridjis, Homero, 242
Arizona, 181–96, 197–209, 210–22
  Border Patrol in, 206, 219, 227–28
  Brown Canyon in, 212, 213–14
  Buenos Aires NWR in, 212–14
  cacti in, 220–21
  Colorado River in, 182–88
  Coronado National Forest in, 197, 202–3
  Guadalupe Canyon in, 206–9
  Little Colorado River in, 187–88
  Meteor Crater in, 189–90
  monarch migration in, 196, 200–201, 202, 207–8
  Organ Pipe National Monument in, 218–22, 273
  Pinery Canyon in, 197–201
  Salt River in, 191–95
  San Carlos River in, 195–96
  San Pedro River in, 210–11, 235
  Sunset Crater in, 188, 189

Art, monarchs depicted in, 243
Asclepiadaceae family (milkweeds)
  and Apocynaceae family, 26
  Asclepias fascicularis, 7, 98–99
  Asclepias speciosa, 7, 22
  and Muellerian mimicry, 36–37
Asian monarch butterflies, 133, 230
Aspens, golden leaves of, 175
Australia, monarch migration in, 247

Babbitt, Bruce, 285
Bagdonas, Karölis, 164
Baker, Robin, 58
Balsamroot, 40 243
Barringer, D. M., 189–90
Bateman Island, 83–85
Bates, H. W., 66
Batesian mimicry, 66
Bats, leaf-nosed, 221–22
Beach Boys, 242
Bear River, 151, 164
Bear River Migratory Bird Refuge, 145, 149–51
Bearings, vanishing, 13, 172
Beaver River, 173–76
Bees, as pollinators, 80–81, 130–32
Beetles
  blister (meloid), 44, 194
  carrion, 157–58, 176
  cicindelid, 176
  colors of, 36–37
  longhorn, 36, 87
  metallicism in, 78
  microtransmitter tracking of, 6
Behavior
  local traits in, 48–49
  variable, 45–46
Bell, Christopher, 243
Big Flat, Utah, 175–76
Biggs, Oregon, 100, 101
Billingsley Creek, 136
Birds, 58, 89–90, 113. *See also* Predators
Birdwings, 29–30
Black Rock Desert, Nevada, 258–60
Black Rock Desert, Utah, 171–72
Blane, Rob and Kathy, 141
Blues, 134, 172, 212
  acmon, 53, 107
  ceraunus, 207, 231
  counting scales of, 34
  eastern tailed, 53, 54
  marine, 184, 207
  parallel evolution of, 53–54
  pygmy, 132, 133, 207
  solitary, 207
Boats, migration on, 54–55
Bonneville Salt Flats, 160–67, 177
Borkin, Susan, 27
Branch, David, 64–65, 66–72, 126, 149, 195, 270
Brewer, Jo, 148, 266
Britain, 54, 96, 150, 219
British Columbia, 19–27, 28–35
  Butterfly World in, 28–30
  milkweeds in, 22, 25, 26
  monarchs in, 21–26, 34, 35
  northernmost monarch breeding grounds in, 14
British Museum of Natural History, 147
Brower, Andrew, 273
Brower, David, 185
Brower, Jane van Zandt, 66
Brower, Lincoln Pierson, 9, 207

on milkweeds, 32, 50
on mimicry, 66
on monarch protection, 246, 286
on predation, 77–78, 241–42
Brown, Rick, 235
Brown Canyon, 212, 213–14
Browns, red-bordered, 199, 200
Brugger, Kenneth and Cathy, 286
Brush-footed butterflies, 85
Buchmann, Stephen, 80–81
Buckeyes, 29, 254
Buenos Aires NWR, 212–14
Bullock-Webster, Julia Rachel Price, 26
Butterflies
accidental migration by, 54–55
basking on ground, 161
bricking of, 12–13
in classrooms, 287
colors of, 23, 32, 79
as diurnal, 58
farming of, 28–30
gardening for, 143, 145, 195
hilltopping by, 116
host plants located by, 39, 98
life expectancy of, 4, 6
metallicism in, 78–79
metamorphosis of, 3
mimicry of, 65–66, 193, 199
mixing genetic stocks of, 286–87
multibrooded, 195
names of, 147–48
palatability spectrum of, 35, 241
pesticides and, 15, 35–36, 128
playing possum, 39, 45, 48, 61
protection of, 9, 277–80
range expansion of, 6, 56
on road kill, 117
scales of, 23, 34
transfer and release of, 286–87
wings of, 23
winter diapause of, 6
See also Caterpillars; Migration; Survival; *type of butterfly*
Butterfly Town, U.S.A., 245
Butterfly World, Kelowna, B.C., 28–30

Cabeza Prieta NWR, 216–17
Cache Valley, Utah, 152, 154
Cacti, 220–21
Caddis flies, 75
Calaveras Grove of Big Trees, 255–56
California, 236–52, 253–63
Central Valley in, 240, 254, 271–72
Cibola NWR in, 228–35
coast of, 240, 249, 272
conservation in, 9, 245–47, 285
Friends of the Monarchs in, 245
as migration destination, 7, 8, 49, 101–2, 139, 207, 245, 269–73
Mojave Desert in, 236–39
monarch migration in, 3, 201, 249, 250–51
monarch parades in, 243, 245
Monarch Program in, 9, 179, 246, 270, 272
monarchs in history of, 244–45
monarchs overwintering in, 7, 70, 245, 250, 272, 286
Morro Bay, 240–46, 248–52, 263
Native Plant Society of, 246–47
Natural Diversity Data Base of, 246

California (*cont.*)
recovery in Arizona of butterflies from, 201
recovery of Idaho butterflies in, 139, 272
recovery of No. 09727 in, 270–73
recovery of transplants in, 138–39
roost trees in, 70, 241, 245, 246–47
Calvert, Bill, 178
Canada. *See* British Columbia
Canada geese, 66–67, 149, 178, 208
Carney, Dan, 86
Carr, Rosie and Cecil, 55
Carson, Rachel, 185, 243
Cascade Range, 102, 108, 121
Caterpillars
frass of, 21
monarch, 3, 31, 33–34, 35, 37–38, 40, 74
for movies, 63
parasites on, 41, 152
in tequila or mescal bottles, 203
tubercles of, 31, 36
walkabouts by, 64
woolly bear, 37, 85, 96, 144
Cattle
coexisting with wildlife, 209
grazing of, 44–45, 112, 120, 121, 165, 176
terracettes of, 121
Cayucos, California, 251–52
Checkerspots, 116
Chelan County, Washington, 11–14, 43
first monarchs recorded in, 11–12, 40
Chiricahua Mountains, 197–98, 203, 208
Chopacka Mountain, 18
Chrysalides, 3, 78, 262–63
Chrysalis, defined, 78
Cibola NWR, 228–35
City of Rocks Natural Preserve, 145–46, 148
Classrooms, butterflies in, 287
Clear Lake Waterfowl Management Area, 168–71, 176
Climate, 88, 110, 250
Clock-shifting, 57–58
Cody, Robin, 124
Coleville Indian Reservation, 52
Collectors
precise labeling needed by, 20
range extension verified by, 62
restrictions on, 183, 229
species named by, 147–48
Colonies
sizes of monarch, 244
See also specific colony
Color vision, 170–71
Colorado Plateau, 180, 185
Colorado River, 182–88
Cibola NWR on, 228–35
crossing Mexican border, 226–27
damming of, 185–86
first navigation of, 185
Glen Canyon of, 183–87
Grand Canyon of, 179, 183, 186–88
Green River and, 179, 235
jetskis on, 231
as monarch conduit, 182, 234–35
Salt River and, 191

Colorado River Compact (1922), 186
Colors, of monarchs, 32
Columbia Basin Project, 57
Columbia Gorge Scenic Area, 99, 270
Columbia Park, 86–87
Columbia River, 43–56, 73–100
  Bateman Island in, 83–85
  butterfly studies near, 14–15
  chemical runoff into, 76
  coulees and, 43, 46, 63
  damming of, 15, 43, 51–52, 55, 57, 74, 102
  ferries across, 52–53, 54
  first salmon released in, 13
  Gallagher Flat on, 11–14, 40
  Grand Coulee area of, 46–55
  Hanford Reach on, 76, 77–82
  jetskis on, 124
  John Day Dam on, 97–99
  McNary Dam on, 94–95
  Methow River and, 14–15
  Okanogan River and, 15, 40
  Snake River and, 82, 87, 90–92, 156
  Tri-Cities area of, 82–85
  and Two Rivers Park, 84, 87, 92
  undammed stretch of, 76, 77
  as Washington-Oregon border, 94
Conservation, of monarchs, 9, 245–47, 285–88
Continental Divide, as migration barrier, 8–9, 139, 140, 207, 269–70
Coppers, 15, 33, 60, 62, 121, 172
  parallel evolution of, 53–54
  purplish, 14, 31, 115
Coral Pink Sand Dunes, 179–80
Coronado, Francisco Vásquez de, 166, 235
Coronado National Forest, 197, 202–3
Coulees, 43, 46–49, 63
Covered bridges, 25–26
Cox, Shirley, 202
Crab Creek, 59–72
  Lower Crab Creek Wildlife Area, 67–69
  monarch tagged near, 68–72
Crescents, 33, 156, 172
  field, 61–62
  mylitta, 14, 61, 122, 151
  pearl, 24
Crows (milkweed butterflies), 78
Dams
  removal of, 77
  and salmon, 77, 102
  See also Habitat; *specific river*
Damselflies and dragonflies, 79
Danaiines (milkweed butterflies), 80
Danaus
  chrysippus, 133, 230
  gilippus berenice, 192–93, 230
  naming of, 148
  See also Monarch butterflies
Darwin, Charles, 66
Davis, Don, 147
Davis Canyon, 39
Day, John (Mah Hah), 108, 115
Day of the Dead, 127, 129
Delores (Papago elder), 215–17
Deming, Alison, 210, 242
Deschutes River, 101–8
Deseret, meanings of, 165

Desert broom, 207
Desert horned lizard, 179–80
Deutsch, Barbara, 195
Diapause, winter, 6, 42, 115
Dixie Summit, Oregon, 115–16, 125
DNA analyses, 49, 274
Dogbanes, 26–27
Donner Pass, 163–64
Douglas County, Washington, first monarch recorded in, 40
Doug's Beach, Washington, 270
Dragonflies, 79, 222, 231
Dreams, monarchs in, 59–60, 105, 149, 166–67, 219–20, 240–41
Drug Enforcement Agency, 206
Dry Falls, Grand Coulee, 51
Duck Valley, Nevada, 156–57
Dugway Proving Grounds, Utah, 164
Dune buggies, 166, 180
Dwarf yellows, 87, 124–25

Eldorado, 166, 235
England. *See* Britain
Ensor, Patti, 87
Eucalyptus groves, 70, 245, 246–47
Evans, W. H., 203
Evolution
  of local traits, 48–49
  of migration, 115
  mimicry in, 65–66, 199
  parallel, 39, 53–54
  random dispersal in, 95–96
  variable behavior in, 45–46
  See also Survival
Extinction, and land use, 219

Fandango Creek, 261–63
Ferguson, Denzel and Nancy, 120
Fire, benefits and threats of, 256
Flicker fusion frequency, 170
Frass, 21
Frémont, John C., 136, 150–51, 172, 237
Friends of the Monarchs, 245
Fritillaries
  coronis, 18, 54, 107, 120
  Diana, 193
  great-spangled, 21
  gulf, 211
  Leto, 19, 21, 121
  male vs. female emergence of, 80
  metallicism in, 78–79
  Nokomis, 160
  regal, 287
  variegated, 190, 207
Frost, Robert, 93
Funk, uses of word, 103

Gallagher Flat, Washington, 11–14, 40
Gender, of monarchs, 80
George, Jean Craighead, 48
Gibo, David, 154–55, 162
Gila monsters, 214–15
Gila River, 196, 235
Gill, Rob and Eve, 273
Glen Canyon, 183–87
Gliding behavior, 3, 51, 154–55, 162, 243
Gloyd Seeps Wildlife Area, 60–62
Glycosides, cardiac, 31–32, 66, 77
Grand Canyon, 179, 183, 186–88, 235
Grand Coulee, 46–55
  and Columbia River, 51–52, 55
  Dry Falls of, 51

flooded towns of, 55
monarch tagged near, 48–50, 61
and Sun Lakes State Park, 47–51
Grand Tetons, Snake headwaters in, 142, 144
Grande Ronde lava flows, 45
Grande Ronde River, 124–25, 126
Great Basin, 132, 156, 159–80, 273
Black Rock Desert of, 171–72
Bonneville Flats in, 160–67
Clear Lake area of, 168–71, 176
and Donner Pass, 163–64
and Great Salt Lake, 164
monarchs in, 161–62, 166, 170, 171
naming of, 172
Sunstone Knoll in, 167–68
White Sand Dunes in, 165–66
Great Britain, 54, 96, 150, 219
Great Salt Lake, 150–52, 160, 164
Great Salt Lake Desert, 177
Green River, 179, 182, 235
Guadalupe Canyon, 206–9, 235
Guzman Loera, Joaquin, 206

Habitats
conservation and protection of, 30, 185, 245–46, 285–88
degradation and destruction of, 15, 30, 45, 52, 55, 76, 77, 81, 185, 218–19, 245, 286, 287
edge of range, 104
fidelity to, 18
mimicry and overlap of, 199, 201
and range expansion, 88
sustainable agriculture in, 218–19
as unofficial countryside, 233
Hackberry butterflies, 108, 194–95, 207, 213
Hairstreaks, 54, 201, 207, 213, 214
Hadley, Drum, 209
Halophytes, 132–33
Hanford Nuclear Reservation, 76, 81
Hanford Reach, 76, 77–82
Hanson, Roseann and Jonathan, 211–14
Hell's Canyon, 118, 122–26, 133
Hellstern, Ron, 152
Helms, Natalie, 270
Hemingway, Ernest, 140–41
Hemp, Indian, 26–27
Henshall, Mary, 139, 152, 234, 272, 287
Herbicides, 35–36, 286
Hewes, Laurence Isley, 196, 198
Hibernation, 6, 24–25, 42, 85, 86, 115
Hill, James, 105
Hill, Sam, 99–100
Hilltopping, 116
Hinchliff, John, 101
Hinkley family, 173
Holland, W. J., 199
Hoover, Herbert, 188
Hopfinger, John, 15, 18, 26, 36
Hornworms, 63
Host plants
herbicides and, 36, 286
location and selection of, 39, 98
and unpalatability, 31–32, 35, 66
See also specific plant
Houck, Mike, 251
Hunter's butterflies, 104

Idaho, 129–44
migration studies in, 138–40, 272
Snake River in, 129–38, 142–44, 155–57

Indians
- ancient trails of, 104–5
- Coleville, 52, 53, 54
- and depopulation, 218–19
- and Deschutes River, 106
- and Geronimo Surrender Memorial, 204
- Paiutes, 157, 174
- petroglyphs of, 97, 102, 107
- pictographs of, 113
- reservation casinos of, 102
- sacred poles of, 13
- and salmon, 13, 106–7, 136
- Tohono O'odham, 215–18
- treaty rights of, 106–7
- Warm Springs, 106
- Yakama, 106

Irrigon, Oregon, 96
IUCN (International Union for Conservation of Nature and Natural Resources), 246

Jackson, Ned, 145–46
Janovy, John, Jr., 235, 258
Jeffers, Robinson, 262
John Day Dam, 97–99
John Day Fossil Beds, 111, 113
John Day River, 101, 102, 108–17
Johnson, Robert, 243
Joshua trees, 237
Journey North, 9
Julia, Mexico, monarch site in, 243–44

Kelowna, British Columbia, 28–32
Kerb, Barbara and Richard, 145
Keremeos Grist Mill, 26
Kinnell, Galway, 242
Kittredge, Bill, 72
Klimt, Gustav, 148
Kluk, Krzysztof, 148
Knight, Helen, 28
Kohler, Steve, 135

Ladies, 152
- Virginia, 207, 212–13
- west coast, 61, 87, 137, 143, 156, 252
- *See also* Hunter's butterflies; Painted ladies

Lake Bonneville, 163, 177
Lake Lahontan, 259
Lake Powell, 185
Lake Roosevelt, 55
Lakes, ancient, 163, 177, 259
Lane, John, 245
Lanting, Franz, 243
Larvae. *See* Caterpillars
Lepp, George, 243
Lewis and Clark Expedition, 85, 92–93, 121
Life expectancy, 4, 98
Linnaeus, Carolus, 148
Lishman, William, 178
Literature, monarchs in, 242–43
Little Colorado River, 187–88
Loiseau, Ron, 21, 24
Lopez, Barry, 171, 258, 259
Lovenfosse, Jeremy, 270, 271, 272
Lower Columbia Basin Audubon Society, 86
Lunar birdwatching, 113
Lunar eclipse, 112–13

Mabey, Richard, 233
McFall Hotel (Shoshone), 140–41

McNary Dam, 94–95
Magnetic fields, orientation via, 8, 58–59
Malpai Borderlands Group, 209
Manhattan Project, 76, 81
Marble Canyon, 182
Marblewings, 34, 40
Marriott, David, 246, 273
Marsha (butterfly net), 6, 13
Martin, Leon and Anne, 135
Maryhill, Washington, 99–100
Mating behavior, 2, 22, 194, 247–48
Meconium, 70
Mennell family, 19, 20–24, 25
Metabolism, in hibernation, 42
Metallicism, 78–79
Metalmarks
  Mormon, 14, 26, 207
  Palmer's, 211
Metamorphosis, of monarchs, 3
Meteor Crater, Arizona, 189–90
Methow River, 14–15
Mexico
  conservation in, 9, 246, 285–86
  Day of the Dead in, 127, 129
  Julia monarch colony in, 243–44
  as migration destination, 7, 8, 49, 207–8, 212, 262, 269, 273
  monarch migration in, 1–4, 201, 202
  monarch predation in, 77–78, 241–42
  monarchs crossing border of, 207–8, 211–12, 273
  monarchs at streamsides in, 249–50
  Pyle's border crossings to and from, 212, 222, 223–27
  See also Michoacán, Mexico
Michoacán, Mexico, 127, 175, 287
  logging in, 287
  magnetized highlands of, 58
  monarch migration into, 233–34
  monarchs overwintering in, 1, 7, 151, 250
  northward monarch migration from, 1–4
  predation on massed monarchs in, 77–78, 241–42
Microtransmitters, 6
Migraine sparkles, 174–75
Migration, 1–9
  accidental, 54–55, 96
  arrival at terminus, 233–34, 244
  in Australia, 247
  beginning of, 1–4
  behavior of butterflies in, 2–3
  of birds, 58
  on boats, 54–55
  chemical barriers to, 15, 36, 286
  destinations in, 7, 8, 49, 101–2, 139, 140, 207–8, 245, 262, 269–73
  distances covered in, 271
  DNA analyses and, 274
  evolution of, 115
  fall, 2, 4, 7, 24, 50, 145, 249
  gliding strategies in, 154–55
  ground observation of, 5–6, 8, 41, 172
  historical record of, 50–51
  of large vs. small butterflies, 124
  of males vs. females, 80
  and mimicry, 199, 201

Migration (*cont.*)
of mixed species, 23–24, 125, 195
and mortality, 250
mystery of, 9, 144, 239, 262
navigation in, 7, 8, 57–59, 160, 269
nectar as limiting factor in, 58, 81, 172, 286
northern limits of, 14, 266
of overwintering species, 115
physical achievement of, 7–8
protecting routes of, 9, 285–88
random dispersal in, 95–96, 172
and range expansion, 6, 56, 124–25
research about, 9, 57–59, 138–40, 273–74, 285, 287
roads as corridors of, 36
Rockies as barrier to, 8–9, 139, 140, 207, 269–70
seasons and, 132
spring, 1–4, 7, 24, 50
of transplanted butterflies, 8, 138–39, 287
water ballasting in, 154–55
western vs. eastern paths of, 8–9, 49, 101–2, 139, 140, 207, 269–70
See also Watercourses; *specific river or type of butterfly*
Milkweed butterflies. *See specific milkweed feeders*
Milkweeds
Asclepias species of, 7, 22, 98–99
bitter latex of, 3, 31–32, 35, 74
in British Columbia, 22, 25, 26
expansion of, 15
herbicides and, 36, 286
historical records of, 50
as host plant, 2–3, 7, 22, 32, 36, 74
human consumption of, 74
resilience of, 287
similarity to dogbanes of, 26
Miller, Grace Phelps (Pyle's grandmother), 43
Miller, Minnie, 136–38
Mimicry, 65–66, 193, 231
Batesian, 66
Muellerian, 36–37, 66
range overlap in, 199, 201
of snakes, 168
and unpalatability, 66, 199
Mojave Desert, 236–39
Monarch, coinage of term, 147
Monarch butterflies
atlas of occurrences of, 101
center of gravity in, 154–55
DNA studies of, 49, 274
in dreams vs. reality, 59–60, 105, 149, 166–67, 219–20, 240–41
eggs deposited by, 2–3, 138
fish patches of, 97, 99, 100
freedom of, 49
gender of, 80
grace of, 41
host plants for, 2–3, 7, 22, 31–32, 36, 74
insights learned from, 269–70
landowners' guide for, 246
in literature and art, 242–43
metamorphosis of, 3
migration of, 1–4, 7, 24, 50, 145, 249
names for, 4, 6, 127, 147, 215
northernmost range of, 14, 266

palatability spectrum of, 35, 241
parasites of, 41
as pollinators, 81
protection of, 9, 245–47, 285–88
reasons for fascination with, 6–9
research about, 49, 57–59, 245–46, 273–74, 285
species similar to, 16, 61, 65–66, 79–80, 192–93, 194, 198, 199, 213, 224
unpalatability of, 31–32, 35, 44, 65–66, 77–78, 199, 241–42
vision of, 170–71
wing loading by, 154–55
See also Caterpillars; Tagging; *specific butterfly or plant*
Monarch Program (San Diego), 9, 179, 246, 270, 272
Monarch Project (Xerces Society), 245–46
Monarch Watch (Kansas), 9, 57–59, 130
Monasterio, Fernando Zorro, 58
Monroe, Mía, 179, 235, 246, 253
Moon, eclipse of, 112–13
Mormons, 152, 164–65, 172, 173
Morro Bay, California, monarch colony in, 240–46, 248–52, 263
Mortality
and migration, 250
See also Aging; Extinction
Moses Coulee, 43–44, 45, 51
Moths
buck, 119–20
giant silk, 285
grass, 134, 163
Isabella, 37
luna, 287
miller, 134, 163
Polyphemus, 24
red underwing, 67–68
sesiid clearwing, 24
silver-marked, 74
tiger, 36, 37, 74, 95, 134
Mourning cloaks, 19, 83, 110–11, 120, 121
hibernation of, 6, 42, 115
Mueller, F., 66
Muellerian mimicry, 36–37, 66
Muir, John, 255
Muir Woods, 246, 253
Myers, David, 114

Nabhan, Gary, 215–18, 219, 232–33
and *Desert Smells Like Rain,* 203–4, 218
in Grand Canyon, 182, 187, 235
on pollination, 80–81
Nabokov, Vladimir
in Arizona, 203
on butterflies, 5, 16, 34, 70, 200, 203, 242, 248
and Véra, 16, 203
Names, scientific, 147–48
National Geographic, 196, 198
National Park Service
A'al Waipia depopulated by, 218–19
and cattle grazing, 112
Nature Conservancy, The, 39, 136–37, 246
Navigation, in migration, 7, 8, 57–59, 160, 269

Nectar
  choice of, 39–40, 47
  learning location of, 98
  in migration patterns, 58, 81, 172, 286
  need for, 50
  See also specific plant
Nevada, 156–59, 258–60
New Mexico, 203–4
Newell, Gordon, 243
Nez, Lewis, 186–87
Nichols, Audley Dean, 205
Nighthawk, Washington, 17–18, 19
North Roosevelt, Washington, 97–99, 270

Oasis theory, 160, 162–63
O'Brien, Chris and Mike, 137, 138
Odonates, 79
Okanagan Valley, British Columbia, 28–35
Okanogan, Washington, 28, 42
Okanogan River, 15, 35–42, 55
Ommatidium, 170
One Thousand to One, 22
Oranges
  sleepy, 195, 199, 207
  tailed, 207
Orange-tips, Pima, 196, 198
Oregon, 12, 263–67
  Columbia River in, 94–97, 100
  Deschutes River in, 101–8
  John Day River in, 101, 102, 108–17
  Snake River in, 118–26, 127–29
Oregon Trail, 109, 117, 263
Organ Pipe National Monument, 218–22, 273
Osoyoos Lake State Park, 15–17
Otters, river, 134
Owl butterflies, 29
Owyhee River, 157, 158

Pacific Grove, California, 243, 245
Paetzel, Mary, 265
Painted beauties, 207, 212–13
Painted ladies, 60–61, 90, 143, 152, 212, 230, 231, 232, 233
  meconium of, 70
  migration of, 6, 23–24, 195
  names for, 156
Paiute Indians, 157, 174
Papua New Guinea, 29–30
Parnassian butterflies, 6, 148
Pentilla, Dan, 97
Perez, Sandra, 57, 58–59
Pesticides, 15, 35–36, 128, 286
Peterson, Roger Tory, 285, 287
Petiver, James, 147–48
Pheromones, 80, 194
Pinery Canyon, 197–201
Polymorphism, balanced, 46
Portal, Arizona, 200–201, 202, 203
Potholes, 43, 63–64, 67, 76
Powder River, 121–22, 126
Powdermilk (Honda), 6, 116, 153, 239, 266, 268
Powell, John Wesley, 185
Predators
  ambush bugs, 48, 107
  blue jays, 65–66, 77
  grosbeaks, 77, 241–42
  mimicry and, 36–37, 199
  orioles, 77–78, 241–42
  other birds, 242
  praying mantis, 35
  swallows, 77, 90, 161–62

unpalatability and, 32, 65–66, 77–78, 241–42
Priest Rapids Dam, 74–75
Purples, red-spotted, 65, 193
Pyle, Bud (Howard, brother), 13, 121–22, 124
Pyle, Robert Campbell (grandfather), 189
Pyle, Thea Linnaea, 32, 35–36, 114, 228, 268, 269
in British Columbia, 25, 26, 29
monarchs first recorded in Chelan by, 11–12, 40
in Utah, 160
in Washington, 11, 14, 17, 18, 37–38, 47–48, 52

Queen butterflies, 195, 207, 209, 211, 224, 230, 231, 232
comparison of monarchs and, 192–93, 194, 213
mating of, 194
migration of, 193, 213

Rabbit, pygmy, 44
Rabbitbrush, 39, 43–44, 93, 96, 133–34, 152, 157, 170, 172, 237
Railroads, 140–41, 154
Reed, Lisa, 88
Riley, C. V., 247–48
Ringlets, 121, 156
ochre, 60, 108, 110, 122
Ritter Island, Idaho, 136–38
Rivers. *See* Watercourses; *specific river*
Road-killed creatures, 117, 153, 157–58, 203, 224, 249
Roads, as migration corridors, 36
Rock climbers, 146
Rocky Mountains, as migration barrier, 8–9, 139, 140, 207, 269–70
Roosevelt, Franklin D., 51
Roosevelt, Theodore, 142
Russian olive trees, 70, 83, 137, 160

Sacajawea State Park, 90, 92, 126
St. John, Dennis, 34–35
Sakai, Walter, 241, 246
Salmon
dams and, 77, 102
Indians and, 13, 106–7, 136
population collapse of, 52
Salt River, 191–95, 235
San Carlos River, 195–96
San Pedro River, 210–11, 235
Sanpoil River, 53
Satyrs, 199–200, 203
Scudder, Samuel, 147
Seasons, and migration, 132
Sequoias, 255–56
Sevier River Valley, 176–78, 180
Sewage farms, 89–90
She Who Watches petroglyph, 102
Shepard, Jon, 16
Shepardson, Lucia, 244–45
Sierra Club, 185
Similkameen River, 17–26
Sisters
Arizona, 197–98, 202, 213
California, 254
Size of butterflies, 124
Skaha Lake, 33–34
Skippers, 14, 40, 83, 116, 121, 156
black, 207
brigadier, 203
common checkered, 40

Skippers (*cont.*)
- fiery, 254
- giant, 203
- great white, 81, 133–34
- Juba, 39, 66, 122
- long-tailed, 207
- sachem, 86, 88–89
- sandhill, 86
- sooty-wing, 254
- stigmata of, 86
- tawny, 86
- woodland, 33, 40, 48, 54–55, 86, 88, 112
- Yuma, 48

Slaughter, John, 208
Smith, Nancy Klos, 243
Smyth, Maurita, 88
Snake River, 118–26, 127–38, 142–44
- Big Bar in, 122–23, 126
- at Columbia River, 82, 87, 90–92, 156
- damming of, 134, 142
- Farewell Bend in, 117, 118, 126, 127
- headwaters of, 142, 144
- Hell's Canyon of, 118, 122–26, 133
- jet-boats on, 123–24, 144
- as monarch conduit, 90–91, 92, 114, 120, 124, 126, 134–35, 138, 143, 156, 158, 159, 164
- Ritter Island in, 136–38
- river otters in canal near, 134
- Three Island Crossing on, 155–57
- and Two Rivers Park, 84, 87, 92

Snake River Birds of Prey National Conservation Area, 133–34
Snake River Plain, 117, 133, 140, 143, 145, 154, 159, 273
Snakes
- gopher, 168, 202, 208
- other species, 202

Snout butterflies, 195, 207, 211, 214, 232
Snow, Katrin, 155, 246
Snyder, Gary, 171, 258
Spanish fly, 44, 194
Spokane River, 126
Squirrels
- Abert's (tassel-eared), 181, 188
- Kaibab, 181–82
- Mount Graham red, 196
- red, 173
- Western gray, 249

Steinbeck, John, 242–43, 247
Steinhart, Peter, 215–17
Sulphurs, 14, 48, 61, 79, 90, 107, 121, 172, 212, 213
- common, 33, 156, 176, 190
- dainty, 87, 124–25, 126, 200, 201, 207
- dogface, 201, 207
- giant cloudless, 199
- hybridization of, 130
- orange, 39, 109, 130, 151, 156, 190
- range expansion of, 201
- yellow, 109, 120, 130

Sun compass, 8, 57–59, 269
Sun Lakes State Park, 47–51
Sunflowers, 46–47, 48, 49, 50, 122–23, 152
Suni, Francisco Chico Tulpo, 216–17
Survival, 18, 90, 98, 115, 250
- adaptive shifts for, 70, 245, 247
- bull's-eyes in, 53

metallicism for, 79
and playing possum, 39, 45, 48, 61
random dispersal for, 95–96
startling predators for, 67–68
tails sacrificed for, 14, 53
unpalatability for, 31–32, 65, 77
See also Winter survival
Sutherland, Faye (Monarch Lady), 139, 272, 287
Swallowtails, 116, 193
anise, 46, 63, 160, 254
in Butterfly World, 28–29, 30
giant, 213
Kaibab, 183
Oregon, 12, 14, 19–20, 26
pipevine, 65–66, 193, 195, 207, 211
tiger, 16, 18, 19
western tiger, 24
Swans, hunting of, 150

Tagging, 13, 250
at Brown Canyon, 214
in Cayucos, 251–52
at Columbia Park, 86–87
at Hanford Reach, 77, 78, 79–80, 82
in Hell's Canyon, 122
at Irrigon, 96
in Monarch Program, 246
No. (09726), 270
No. (09727), 270–73
No. (81726), 21–23
No. (81741), 48–50, 61
No. (81746), 68–72
No. (81783), 90, 91, 92
No. (81784), 90, 92
No. (81792), 177
No. (81994), 214
No. (81999), 252
No. (82000), 252
at North Roosevelt, 97–98
near Priest Rapids Dam, 74–75
process of, 22–23
purpose of, 22
and recovery, 8, 9, 138–39, 270–73
at Salt River, 194
sample tag for, 22
and splinting damaged wing, 100
transfers, and release, 138–39
Tamaracks, 256
Tamarind, 256
Tamarisk, 132, 133, 256
Taylor, Orley, 130
Teale, Edwin Way, 144, 153, 160, 161, 163
Temperature, 110, 250
Thistles, as host plants, 24, 60, 156
Thousand Springs, 136
Three Island Crossing State Park, 155–57
Tintic Valley Project, 165
Tortoiseshells
California, 121
Compton, 55
impact of pesticides on, 35–36
Milbert's, 115, 120
overwintering of, 115
Toxins, in milkweeds, 31–32, 35, 74
Trees
and adaptive shifts, 70, 245, 247
conservation and protection of, 245–47
and oasis theory, 160, 162–63
sap from, 85
See also specific tree

Trichoptera, larvae of, 75
Trimble, Stephen, 170, 258, 260
Tsagagalal petroglyph, 102
Tubercles, 31, 36
Turkmenia, Asian monarchs in, 132–33, 230
Tweit, Susan, 231
Two Rivers Park, 84, 87, 92

Ultralight planes, 6, 41, 178
Umatilla River, 96
Updike, John, 242
Urquhart, Fred
  on barometric pressure, 71
  monarch research by, 59, 138, 144, 196, 286
  tagging program of, 8, 27, 271
Utah, 160–80
  Bear River Migratory Bird Refuge in, 145, 149–51
  Bonneville Salt Flats in, 160–67
  Deseret as name of, 165
  Frémont people in, 172–73
  Great Salt Lake in, 150, 151, 152
  Idaho butterflies' recovery in, 139, 152, 234
  Mormons in, 152, 164–65, 172, 173
  transcontinental railroads in, 154
  Wasatch Front of, 145, 151, 152
  See also Great Basin
UV fluorescence, 170–71

Vaccaro, Ro, 245
Vane-Wright, Dick, 50
Viceroy butterflies, 67, 78, 93, 259
  at Bateman Island, 83, 84
  along Columbia River, 15, 36, 86
  hibernation of, 86
  hybrids of, 124
  as monarch mimics, 65–66, 79–80, 192, 193, 199, 224
  naming of, 147, 148
Vision, of monarchs, 170–71

Walla Walla River, 93–94
Wallula Gap, Washington, 94
Wanderers. *See* Monarch butterflies
Wasatch Front, Utah, 145, 151, 152
Washington, 11–19
  ALE site in, 76
  Columbia River in, 11–15, 43–56, 73–94, 97–100
  Crab Creek in, 59–72
  first dainty sulphurs in, 87, 124
  first sachem skippers in, 88
  Grand Coulee in, 46–55
  herbicides used in, 35–36
  homecoming in, 268–70
  monarchs in, 11–13, 14, 16, 40
  Okanogan River in, 15, 35–42, 55
  Potholes in, 43, 63–64, 67, 76
  Tri-Cities area of, 82–85, 88–89
Wasps, 41, 211, 232
Water tables, 52, 136
Watercourses
  alteration of, 15, 43
  migration along, 43, 59–60, 90–91,
  92, 99, 125, 126, 152, 182, 234–35, 269
  nectar plants along, 79, 125
  preservation of wildness of, 123–24
  and range expansion, 124–25
  See also specific river

Wendell, Idaho, 135
Wenner, Adrian, 56
White Sand Dunes, 165–66, 180
Whites, 18, 61, 90, 121, 137
  Becker's, 54, 156
  cabbage, 31, 33, 67, 81, 83, 95, 112, 133, 143, 224
  Chiricahua pine, 198–99, 200, 201
  pine, 11, 54, 198
  Terloot's, 198–99
  western, 156
Wilbur-Cruce, Eva Antonia, 273
William, Prince of Orange, 147
Williams, C. B., 58, 59
Williams, Terry Tempest, 149, 150
Wilson, Caroline, 210
Wing loading ratio, 154–55
Winter survival, 70, 88
  diapause in, 6
  hibernation and, 6, 24–25, 42, 85, 86, 115
  migration and, 115
  See also Migration
Wood nymphs, 14, 24, 39, 61
Woolly bears, 37, 85, 96, 144
Worley, James, 242

Xerces Society, 245–46

Yakama Indian Reservation, 106
Yakima River, 82–83, 84
Young, Brigham, 165, 177–78
Yucca, 180

Zion, Brion, 165
Zoological Code, 148
Zwinger, Ann, 69, 186

Author photo by Eddie Rivers

ABOUT THE AUTHOR

Robert Michael Pyle holds a Ph.D. from Yale University School of Forestry and Environmental Studies and is the author of eighteen books and hundreds of essays, papers, poems, and stories. He is an internationally known and respected expert on butterflies and their conservation. In 1971 he founded the Xerces Society for Invertebrate Conservation, a nonprofit organization that works to protect pollinators, other invertebrates at risk, and their habitats. Pyle is the recipient of numerous awards and honors for his conservation work and natural history writing, including the Distinguished Service Award from the Society for Conservation Biology. *Wintergreen: Rambles in a Ravaged Land,* which described the devastation caused by unrestrained logging in the Willapa Hills of Washington where Pyle lives, was awarded the 1986 John Burroughs Medal for Distinguished Nature Writing. In 1995 Pyle received a Guggenheim Fellowship to write *Where Bigfoot Walks: Crossing the Dark Divide.* His other books include *The Thunder Tree, Sky Time in Gray's River: Living for Keeps in a Forgotten Place* and *The Butterflies of Cascadia.* In March 2013 Yale University Press published a paperback edition of Pyle's *Mariposa Road: The First Butterfly Year.*